Withdrawn
University of Waterloo

LAKE McILWAINE

MONOGRAPHIAE BIOLOGICAE

VOLUME 49

Editor
J. ILLIES
Schlitz, F.R.G.

Dr W. Junk Publishers   The Hague-Boston-London   1982

LAKE McILWAINE

The Eutrophication and Recovery of a Tropical African Man-Made Lake

Edited by
J. A. THORNTON

With the assistance of
W. K. NDUKU

Dr W. Junk Publishers   The Hague-Boston-London   1982

Distributors:

for the United States and Canada

Kluwer Boston, Inc.
190 Old Derby Street
Hingham, MA 02043
USA

for all other countries

Kluwer Academic Publishers Group
Distribution Center
P.O. Box 322
3300 AH Dordrecht
The Netherlands

Library of Congress Cataloging in Publication Data **CIP**
Main entry under title:

Lake McIlwaine : the eutrophication and recovery of
 a tropical African man-made lake.

 (Monographiae biologicae ; v. 49)
 Bibliography / by Margaret J. Thornton &
J.A. Thornton
 Includes indexes.
 1. Eutrophication--Zimbabwe--McIlwaine, Lake.
2. Limnology--Zimbabwe--McIlwaine, Lake.
3. Water quality management--Zimbabwe--McIlwaine,
Lake. 4. McIlwaine, Lake (Zimbabwe) I. Thornton,
J. A. (Jeffrey A.) II. Series.
QP1.P37 vol. 49 [QH195.Z55]   574s   82-15238
                 [363.7'394'096891]   AACR2

ISBN 90 6193 102 9 (this volume)
ISBN 90 6193 881 3 (series)

Cover design: Max Velthuijs

Copyright © 1982 by Dr W. Junk Publishers, The Hague.

All rights reserved. No part of this publication may be reproduced, stored in a retrieval system,
or transmitted in any form by any means, mechanical, photocopying, recording, or otherwise,
without the prior written permission of the publishers,
Dr W. Junk Publishers, P.O. Box 13713, 2501 ES The Hague, The Netherlands.

PRINTED IN THE NETHERLANDS

The Hunyanipoort Dam and Lake McIlwaine (photo: The Herald, Salisbury)

Preface

> 'And God said, Let there be a firmament in the midst of the waters, and let it divide the waters from the waters.'
>
> *Genesis 1:6*

Lake McIlwaine is a man-made lake. It was formed in 1952 by the Hunyanipoort Dam and is situated on the Hunyani River some 37 km southwest of Salisbury* in the Republic of Zimbabwe**. It is a lake of many aspects: being a popular recreational site, the City's primary water supply reservoir (and the fourth largest impoundment in Zimbabwe), a source of irrigation water to downstream farms, an important fishery, and, until the 1970s, the receptacle of Salisbury's sewage effluent. It is, in short, typical of so many 'urban' lakes in Africa and throughout the world. Lake McIlwaine is also unique, to my knowledge: being amongst the first of the major man-made lakes on the continent to suffer from what is known as cultural eutrophication, and the first to be rehabilitated to a mesotrophic state through a rational programme of lake management. This volume synthesizes this process of eutrophication and recovery in terms of the geology and geography (Chapter 2), physics (Chapter 3), chemistry (Chapter 4) and biology (Chapter 5) of the lake, and, whilst discussion of the trophic relationships between these components is beyond the scope of this monograph, discusses its utilisation, conservation and management (Chapter 6).

* Subsequent to writing, the name of the Zimbabwean capital was changed to Harare on 18 April 1982. – Ed.
** The Colony of Southern Rhodesia was granted independence by H. M. Government on 18 April 1980 and became the Republic of Zimbabwe. All names used in this monograph are those currently in use at the time of writing.

The recovery of Lake McIlwaine was effected through the efforts and foresight of a large numer of individuals: in Government, in the University, in commerce and industry, and in the general public. Their efforts formulated the principles, policies and legislation and, more importantly, implemented these same which resulted in the recovery of the lake. Some of these people have now recorded the roles played by their colleagues and themselves in this endeavour, and my sincere thanks, as editor, go out to them, collectively and individually, for without their expertise this monograph would not have been possible.

Acknowledgement must also be made to the Government of Zimbabwe, particularly to the Ministry of Natural Resources and Water Development; to the City of Salisbury; and to the University of Zimbabwe, who generously funded large portions of the research programmes; and to Lever Brothers (Pvt.) Ltd. and Mobil Oil Zimbabwe (Pvt.) Ltd. and others who equally generously provided monies for various necessary items of equipment. I am grateful for the co-operation and assistance of my colleagues and friends at the University of Zimbabwe, in the Department of National Parks and Wild Life Management, in the Division of Water Development, and latterly in the Hartbeespoort Dam Ecosystem Programme team of the National Institute for Water Research, CSIR, whose advice and encouragement resulted in the completion of this volume.

As editor, I have had to co-ordinate the various sections of this monograph which were written by a number of independent authors and co-authors, and in some instances considerable adjustment was necessary to bring these contributions into line with the general plan of the volume. Should any errors or omissions have occurred during this process, I tender my sincere apologies.

This volume forms contribution No. 4 to the National Water Quality Survey of Zimbabwe.

Pretoria, December 1981   JEFFREY A. THORNTON

Contents

Preface		VII
Addresses of authors		XI

1. **Introduction**
   The creation of Lake McIlwaine: history and design, by N. A. Burke and J. A. Thornton — 1

2. **Geology and geography**
   Land use survey of the Upper Hunyani catchment, by K. Munzwa — 11

3. **Physics**
   Physical limnology, by P. R. B. Ward — 23
   The hydrology of the Lake McIlwaine catchment, by B. R. Ballinger and J. A. Thornton — 34

4. **Chemistry**
   Water chemistry and nutrient budgets, by J. A. Thornton and W. K. Nduku — 43
   The sediments, by R. Chikwanha, W. K. Nduku and J. A. Thornton — 59
   – Sediment chemistry, by J. A. Thornton and W. K. Nduku — 59
   – Sediment transport, by R. Chikwanha — 66
   The effects of urban run-off, by R. S. Hatherly, W. K. Nduku, J. A. Thornton and K. A. Viewing — 71
   – The aqueous phase: nutrients in run-off from small catchments, by J. A. Thornton and W. K. Nduku — 71
   – The solid phase: a study of pollution benchmarks on a granitic terrain, by R. S. Hatherly and K. A. Viewing — 77
   Insecticides in Lake McIlwaine, Zimbabwe, by Yvonne A. Greichus — 94

5. **Biology**
   An SEM study of bacteria and zooplankton food sources in Lake
   McIlwaine, by Monika Boye-Chisholm and R. D. Robarts   101
   Phytoplankton, primary production and nutrient limitation, by
   R. D. Robarts, J. A. Thornton and Colleen J. Watts   106
   – The algal community, by J. A. Thornton   106
   – Primary production of Lake McIlwaine, by R. D. Robarts   110
   – An examination of phytoplankton nutrient limitation in Lake
     McIlwaine and the Hunyani River system, by Colleen J. Watts 117
   Zooplankton and secondary production, by J. A. Thornton and
   Helen J. Taussig   133
   Aquatic macrophytes and *Eichhornia crassipes*, by M. J. F. Jarvis,
   D. S. Mitchell and J. A. Thornton   137
   The benthic fauna of Lake McIlwaine, by B. E. Marshall   144
   The fish of Lake McIlwaine, by B. E. Marshall   156
   Avifauna of Lake McIlwaine, by M. J. F. Jarvis   188

6. **Utilisation, management and conservation**
   Water pollution: perspectives and control, by D. B. Rowe   195
   Water supply and sewage treatment in relation to water quality in
   Lake McIlwaine, by J. McKendrick   202
   Fisheries, by K. L. Cochrane   217
   Recreation, by G. F. T. Child and J. A. Thornton   221
   Research: perspectives, by J. A. Thornton   227

7. **Bibliography**
   by Margaret J. Thornton and J. A. Thornton   233

Taxonomic index   241

General index   245

Addresses of authors

B. R. Ballinger, Hydrological Branch, Division of Water Development, Ministry of Natural Resources and Water Development, Private Bag 7712, Causeway, Salisbury, Zimbabwe.

Monika Boye-Chisholm, 13 Glebe End, nr. Bishops Stortford, Herts., UK (previously: Hydrobiology Research Unit, University of Zimbabwe.

N. A. Burke, Designs Branch, Division of Water Development, Ministry of Natural Resources and Water Development, Private Bag 7712, Causeway, Salisbury, Zimbabwe.

R. Chikwanha, Hydrological Branch, Division of Water Development, Ministry of Natural Resources and Water Development, Private Bag 7712, Causeway, Salisbury, Zimbabwe.

G. F. T. Child, Director, Department of National Parks and Wild Life Management, P. O. Box 8365, Causeway, Salisbury, Zimbabwe.

K. L. Cochrane, Limnology Division, National Institute for Water Research, CSIR, P.O. Box 395, Pretoria 0001, Republic of South Africa (previously: Lake Kariba Fisheries Research Institute, Department of National Parks and Wild Life Management, Zimbabwe).

Yvonne A. Greichus, Enviro Control Inc., 11300 Rockville Pike, Rockville, MD 20852, USA.

R. S. Hatherly, c/o Institute of Mining Research, University of Zimbabwe, P.O. Box MP.167, Mount Pleasant, Salisbury, Zimbabwe.

M. J. F. Jarvis, Branch of Terrestrial Ecology, Department of National Parks and Wild Life Management, P.O. Box 8365, Causeway, Salisbury, Zimbabwe.

B. E. Marshall, Lake Kariba Fisheries Research Institute, Department of National Parks and Wild Life Management, P.O. Box 75, Kariba, Zimbabwe.

J. McKendrick, City Chemist, City of Salisbury, P.O. Box 1583, Salisbury, Zimbabwe.

D. S. Mitchell, Division of Irrigation Research, CSIRO, Private Mail Bag, Griffith, New South Wales, Australia 2680 (previously: Hydrobiology Research Unit, University of Zimbabwe).

K. Munzwa, Institute of Mining Research, University of Zimbabwe, P.O. Box MP.167, Mount Pleasant, Salisbury, Zimbabwe.

W. K. Nduku, Deputy Director, Department of National Parks and Wild Life Management, P.O. Box 8365, Causeway, Salisbury, Zimbabwe (previously: Hydrobiology Research Unit, University of Zimbabwe).

R. D. Robarts, Limnology Division, National Institute for Water Research, CSIR, P.O. Box 395, Pretoria 0001, Republic of South Africa (previously: Hydrobiology Research Unit, University of Zimbabwe).

D. B. Rowe, Central African Power Corporation (CAPCO), P.O. Box 630, Salisbury, Zimbabwe (previously: Planning Branch, Division of Water Development, Ministry of Natural Resources and Water Development, Zimbabwe).

Helen J. Taussig, Division of Nature Conservation, Transvaal Provincial Administration, Private Bag X209, Pretoria 0001, Republic of South Africa (previously: Limnology Division, National Institute for Water Research, CSIR, Republic of South Africa).

J. A. Thornton, Limnology Division, National Institute for Water Research, CSIR, P.O. Box 395, Pretoria 0001, Republic of South Africa (previously: Hydrobiologist, Department of National Parks and Wild Life Management; and Hydrobiology Research Unit, University of Zimbabwe).

Margaret J. Thornton (previously: Ministry of Education and Culture, Zimbabwe).

K. A. Viewing, Institute of Mining Research, University of Zimbabwe, P.O. Box MP.167, Mount Pleasant, Salisbury, Zimbabwe.

P. R. B. Ward, Schultz International Ltd., 1155 West Georgia Street, Vancouver, British Columbia, Canada V6E 3H4 (previously: Department of Civil Engineering, University of Zimbabwe).

Colleen J. Watts, Eulamore Street, Carcoar, New South Wales, Australia 2791 (previously: Hydrobiology Research Unit, University of Zimbabwe).

1 Introduction

The creation of Lake McIlwaine: history and design
N. A. Burke and J. A. Thornton

At the close of the Second World War, a major influx of immigrants into Zimbabwe and the resultant sudden, accelerated expansion of the City of Salisbury, together with a severe drought in 1947–48, resulted in an overtaxing of the City's water resources as supplied from the Prince Edward Dam (Fig. 1). As a result of this water supply problem, the Municipality investigated the possibility of siting a dam further downstream on the Hunyani River, at Hunyanipoort, where the river cuts through the Hunyani Hills (Fig. 1).

Because this site was classed as a national dam site, the Ministry of Natural Resources and Water Development became involved. The City of Salisbury had considered that a dam built to the optimum height allowed by the site would exceed their projected requirements for water in the near future. Government, on the other hand, considered it to be in the national interest that any dam constructed on the site be to the optimum hight allowed. Agreement was reached between the Municipality and Government, and construction of the dam to optimum height was undertaken between 1952 and 1953. The cost of construction and the water to be impounded was to be shared equally between the City and Government. The dam was designed by the Ministry of Natural Resources and Water Development and construction was undertaken by the firms of Clifford Harris and Costain Ltd.

Construction

The geology of the dam site is relatively straightforward, being a fractured banded ironstone exposed at the surface overlaying a solid dolerite outcrop at depth (see K. Munzwa, this volume), and provided a good foundation for the dam (Fig. 2). The cut-off trench was taken down to the level of the dolerite

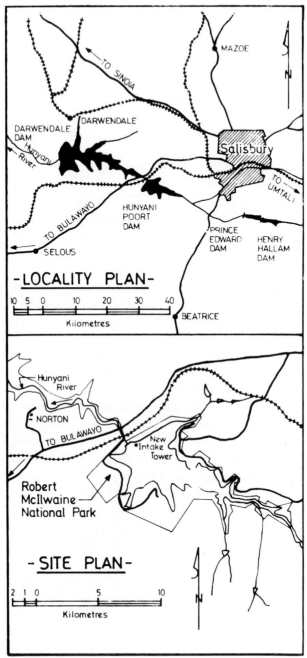

Fig. 1 Location of Lake McIlwaine and the Hunyanipoort Dam in relation to the City of Salisbury, and the situation of the water-works intake tower and public amenities.

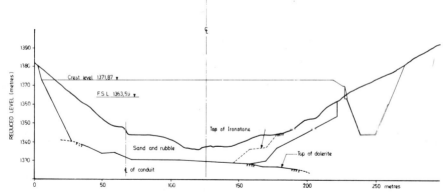

Fig. 2 Section through the Hunyani Poort showing the situation of the dam wall and the general geology of the area.

outcrop on the left bank and through the river bed. On the right bank where the dolerite outcrop swings downstream the cut-off was continued by means of a close centre grout hole curtain. To supplement this grout curtain on the right bank, the right abutment was covered with an impervious blanket connected to the impervious core of the dam, extending from the grout curtain to the spillway cill.

The dam itself is an earth-rock structure with a wide impervious core and a relatively flat upstream slope (Fig. 3). The impervious core material was obtained from within the basin and consisted of an impermeable red sandy clay derived from the ironstone. The rockfill was derived largely from the ironstone obtained during the spillway excavations. This ironstone was found to break up into relatively small-size fractions when handled and compacted, and this was considered to be advantageous as the coarse filter zones were extremely thin. However, the small size of the rockfill allowed beaching to occur on the upstream slopes of the wall as a result of the harsh wave climate created by the prevailing winds which blow straight down the axis of the dam. This situation was recently rectified by the placing of a protective rip-rap on the upstream face.

The spillway (Fig. 4), situated on the right bank, is a side channel design and is fully lined. The structure has stood up extremely well to the large flood flows that have passed through it. The open cut through the hillside above the top of the lining has also proved extremely stable despite the steep unprotected slope (0.7:1).

The original outlet works, situated on the left bank, served both as the river outlet and the outlet to the purification works which were sited just down-

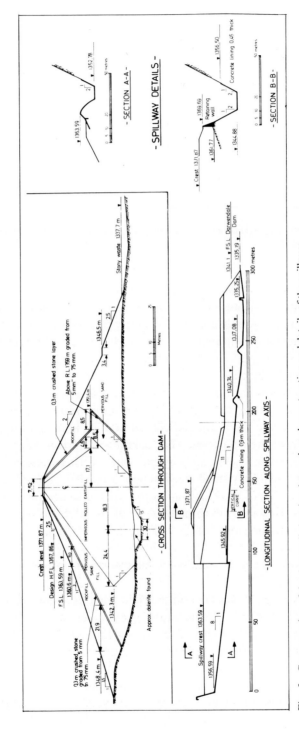

Fig. 3 Cross-section of the dam wall showing the earth-rock construction and details of the spillway.

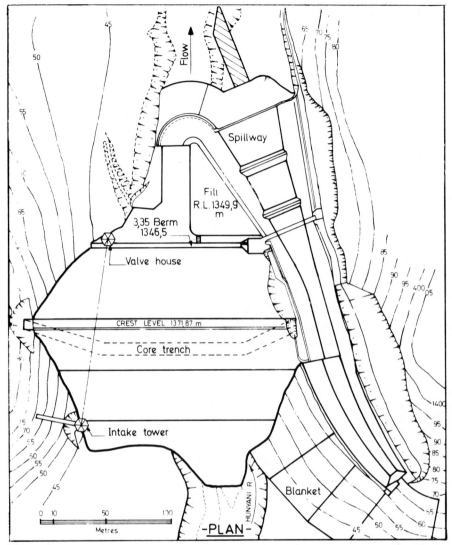

Fig. 4 Plan view of the Hunyanipoort Dam showing the situation of the spillway and the original water-works intake tower.

stream of the wall (Fig. 4). Three inlet levels at 7.5 m, 14.0 m and 18.6 m below full supply level were provided. Due to the prevailing winds preventing circulation of the water in the vicinity of this tower, however, the water supplied to the purification works was of very poor quality (see J. McKendrick, this volume). To rectify this, a new multiple inlet tower was constructed by the City of Salisbury in the basin of the dam upstream of the right bank (Fig. 1).

The completed dam impounded some 250×10^6 m³ of water at full supply level and had a surface area of 26.3 km². A summary of the morphological characteristics of the dam and its basin is given in Tables 1 and 2, and the capacity-surface area curve is shown in Fig. 5.

Table 1 Morphological characteristics of the Hunyanipoort Dam

Characteristic	
Crest level	1371.874 m
High flood level	1367.861 m above mean sea level
Full supply level	1363.594 m
Crest width	7.62 m
Length of wall at crest level	201 m
Spillway crest length	112.78 m
Outflow flood	2518 m³ s⁻¹
1 : 2000 year inflow flood	4233 m³ s⁻¹
Yield	3.11 m³ s⁻¹

Table 2 Morphological characteristics of Lake McIlwaine

Characterstic	
Maximum depth	27.43 m
Mean depth	9.4 m
Maximum breadth	8.0 km
Mean breadth	1.68 km
Length	15.7 km
Shoreline length	74 km
Full supply volume	250×10^6 m³
Full supply surface area	26.30 km²
Catchment area	2227 km²

History

Following the completion of the Hunyanipoort Dam in 1953 and the filling of Lake McIlwaine in the same year, there was an almost immediate biological reaction. Water hyacinth, *Eichhornia crassipes* (Mart.) Solms, which had been present in limited quantities in the Hunyani River system, particularly in the Makabusi River, prior to the construction of the dam, now found the newly created Lake McIlwaine an ideal habitat (see M. J. F. Jarvis, this volume; M. J. F. Jarvis *et al.*, this volume). There was a marked growth of the plant between 1953 and 1962; limited spraying with 2–4 D was begun in 1953 to try to control the spread of the macrophyte. By 1959, the first easily detectable signs of eutrophication were also being noted in the form of algal blooms. The first scientific investigations of the lake were undertaken at about this time (Van der Lingen, 1960), and the City of Salisbury began a more intensive

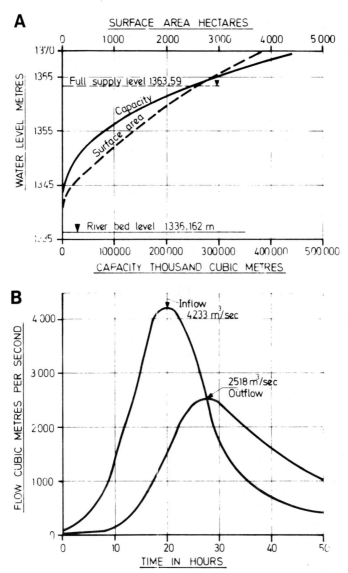

Fig. 5 Capacity-surface area curves and inflow-outflow hydrographs for the Hunyanipoort Dam.

monitoring effort as the effects of eutrophication on their water treatment works became significant (see J. McKendrick, this volume).

In the decade which followed, the growth of water hyacinth remained at fairly low levels with no heavy growths of the macrophyte being apparent during these years. The symptoms of eutrophication, however, continued to appear although by 1964 the lake was no more than mesotrophic (Munro, 1966). Continued input of nutrients and inorganic ions derived largely from treated sewage effluent resulted in the lake being pronounced as eutrophic by the end of the decade (Marshall and Falconer, 1973). Long-term research was initiated by the Hydrobiology Research Unit of the University of Rhodesia (now the University of Zimbabwe) in 1967, initially to determine the causes of the eutrophication of the lake, and latterly to investigate the possibility of recovery following the diversion of nutrients to pasture irrigation schemes. This research effort has now been brought to a conclusion and many of the findings are reported in this volume.

By 1970, there came a massive build up of water hyacinth and extensive deoxygenation of the waters of the lake, epecially at times of overturn, became a common feature, Massive fish kills at this time caused widespread public outcry. The increasing concentrations of nutrients and inorganic ions, combined with the deoxygenation and massive algal and macrophyte blooms, led to the lake being described as hypereutrophic in 1971 (Salisbury Sewerage Disposal Environmental Impact Statement Committee, 1979). Extensive publicity, resulting from the popularity of the lake as a tourist resort and recreational facility (see G. T. F. Child and J. A. Thornton, this volume), brought about the first effective water pollution control legislation at about this time (see D. B. Rowe, this volume), and chemical control of the water hyacinth problem followed shortly thereafter. Thus, by 1972, water hyacinth had been virtually eradicated from the lake surface and municipal wastewater was beginning to be diverted to the irrigation schemes. Effluent diversion continued in stages through to 1977 when nearly 100% of the municipal wastewater was being treated to tertiary standards. The last fish kill was reported in January of 1976, and, although periodic algal blooms still occur, the lake was well on the road to recovery. By the end of the 1970s, Thornton (1980) was able to describe Lake McIlwaine as bordering on mesotrophy.

References

Marshall, B. E. and A. C. Falconer, 1973. Eutrophication of a tropical African impoundment (Lake McIlwaine, Rhodesia). Hydrobiol., 43: 109–124.

Ministry of Natural Resources and Water Development, 1979. Dams in Zimbabwe. Unpublished report, Salisbury.

Ministry of Water Development, 1970. Hunyani Poort Dam. Unpublished leaflet, Salisbury.

Munro, J. L., 1966. A limnological survey of Lake McIlwaine, Rhodesia. Hydrobiol., 28: 181–308.

Salisbury Sewerage Disposal Environmental Impact Statement Committee, 1979. Report on Salisbury's sewerage disposal. MOWD Report, Salisbury.

Thornton, J. A., 1980. Factors influencing the distribution of reactive phosphorus in Lake McIlwaine, Zimbabwe. D.Phil. Diss., University of Zimbabwe.

Van der Lingen, M. I., 1960. Some observations on the limnology of water storage reservoirs and natural lakes in Central Africa. First Fed. Sci. Congress Proc., pp. 1–5.

2 Geology and geography

Land use survey of the Upper Hunyani catchment
K. Munzwa

This study examines the land use patterns within the Upper Hunyani River catchment, an area of 2136 km^2. Its purpose is to facilitate the understanding of the ecological system of Lake McIlwaine at the lower limit of the study area. Lake McIlwaine, together with the other large dams on the Hunyani River, the Prince Edward and Henry Hallam Dams and Lake Robertson (Darwendale Dam), are the main sources of the water supply for both domestic and industrial use in the City of Salisbury, Zimbabwe.

The Hunyani River catchment is predominantly rural and it was believed that there was no real danger to the ecological balance of the catchment area, particularly to those of the four principal dams. However, pollution from industrial and urban wastes has led to an ecological imbalance. Some of its major causes have been examined and others remain to be investigated.

Methods

This study was based mainly on the interpretation of thirty-six aerial photographs. For this reason the land categories, especially those of the vegetation types, are subjective and the boundaries are arbitrary due to distortions caused through compiling the maps using simple photogrammetry techniques. As is almost always the case in any geographical study, the aerial photographs were out of date. Two sets of photographs were used, 1968 and 1977. The 1977 set covered the largest portion of the area and for this reason most of the map is believed to be reliable. The scale of the photographs was 1:80,000. Outlines of the main land use categories were drawn at this scale and were transferred to 1:50,000 maps. This information was compiled and the map appears to be the first to have been made of land use in the Hunyani River catchment (Fig. 1).

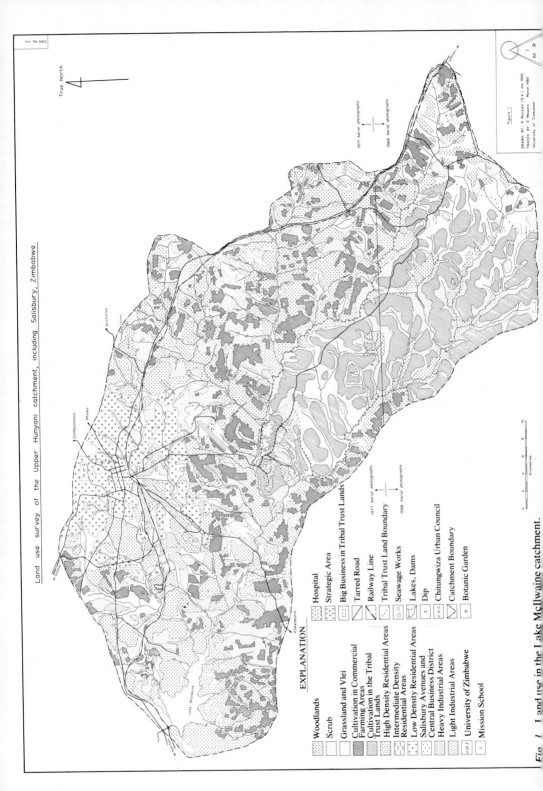

Fig. 1. Land use in the Lake McIlwaine catchment.

Geology

The catchment of the Hunyani River above Lake McIlwaine is underlain by rocks of Archaean age which form a part of the Zimbabwe Basement Complex (Fig. 2). The upper part of the catchment is underlain by rocks of the Older Gneiss Complex and this contains relatively small inclusions of schistose rocks comprised of meta-sediments and meta-volcanics of Bulawayan Age. A relatively small part of the upper extremity of the catchment is underlain by granite.

The lower part of the catchment including Lake McIlwaine and the other large dams of the system is almost entirely underlain by granite except for a relatively large proportion of the northern flank and a part of the northwestern boundary. The rocks beneath the northern flank are comprised of meta-sediments and meta-volcanics also of Bulawayan Age. The City of Salisbury, including its industrial development, extends over almost all of the outcrop and sub-outcrop of these rocks. A narrow belt of schists, including a

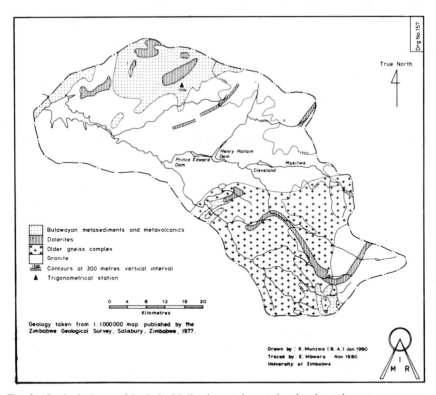

Fig. 2 Geological map of the Lake McIlwaine catchment showing the major water courses.

13

banded iron-stone formation, supports the ridge of hills which form the abutment of Lake McIlwaine (see N. A. Burke and J. A. Thornton, this volume).

There are several dolermite sheets which outcrop throughout the catchment area. These are a part of the Mashonaland Dolermite Swarm and they result in distinctive zones of red soil amongst the much more expansive zones of light coloured sandy soils overlying the granites.

Topography

The Upper Hunyani River catchment is generally a gently undulating featureless plateau (Kay and Smout, 1977). It lies between 1300 and 1500 metres above sea level, with the largest portion of the catchment lying between 1400 and 1500 metres. The southern part of Salisbury extends into the Hunyani River catchment and most of the urban and industrial run-off finds its way into the drainage system close to Lake McIlwaine.

Land use categories

A number of basic land use categories were established, and the results of the investigation indicated their relative importance (Table 1). Two thirds of the catchment is vegetated and two thirds of the remaining third is cultivated. About 10% of the total is affected by urban and industrial development and dams.

Figures 3 and 4 clearly show that vegetation, woods, scrub, and grass covers by far the largest portion of the catchment, an area of 1444 km^2 or 67.6% of the total area. However, this also includes the pasture land in the commercial farming areas. The tribal areas, on the other hand, are almost

Table 1 Land use of the Hunyani River catchment

Land use	Area (km^2)	%
Vegetation cover	1444	67.6
Cultivation	492	23.0
Developed areas (residential/industrial)	166	7.8
Lakes, dams, sewage farms	32	1.5
Other	2	0.1
Total	2136	100.0

devoid of vegetation and as a result are highly susceptible to soil erosion. Various political, social, economic, and physical factors are involved, but the present distribution of the woody vegetation results mainly from the need for fuel-wood and building timber for each family unit in the relatively densely populated, pre-independence Tribal Trust Land areas.

Catchment land use may also be divided in terms of rural and urban categories. Urban refers to the Central Business District (CBD) and avenues, residential townships, and industrial areas, and ignores the official City boundary (Table 2). It is clear that the urban development accounts for a very small proportion of the total area, being only 166 km^2 or 7.8% of the 2136 km^2 catchment. The greatest proportion of the Hunyani River catchment, 92.2%, is rural. This result confirms the belief that the catchment is almost wholly rural. Nevertheless, the level of pollution in Lake McIlwaine has increased alarmingly in the last two decades, primarily because of the location of the reservoir, which lies in the same catchment as the City which it supplies (Kaye and Smout, 1977). Therefore, Salisbury draws its water supply from the same body of water into which its waste products are discharged. These waste products would include urban run-off (see W. K. Nduku and J. A. Thornton, this volume; R. S. Hatherly and K. A. Viewing, this volume), sediments (see R. Chikwanha, this volume), and domestic and industrial effluents (see J. McKendrick, this volume).

Table 2 Land use of the Hunyani River catchment in terms of urban and rural categories

Land use	Area (km^2)	%
Urban	166	7.8
Rural	1970	92.2
Total	2136	100.0

For this study, a more detailed breakdown of land use seemed desirable, and so a total of eleven land use categories were selected for the catchment of the Hunyani. These categories are shown in Fig. 1, and the areas covered by each and their proportions are shown in Table 3. From the table, it is evident that woodlands and scrub account for 43.3% of the total, and that grasslands and vleis amount to about half of that area or 24.2%. Scrubland covers about one half of the woodland area (Figures 3 and 4). Cultivation and farming in the commercial farming areas covers 10.8% of the catchment area and this area is roughly equivalent to the cultivation and subsistence farming in the rural areas which amounts to 12.2% of the total (Fig. 5). The area covered by the

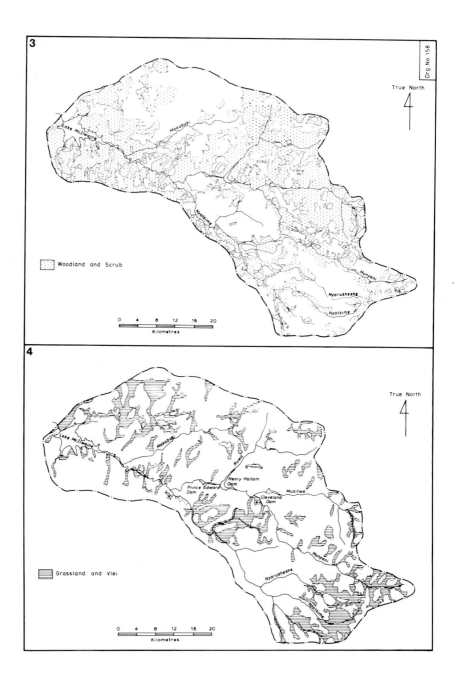

Fig. 3 Distribution of wooded areas in the Lake McIlwaine catchment.

Fig. 4 Distribution of grasslands in the Lake McIlwaine catchment.

Table 3 Eleven categories of land use in the Hunyani River catchment

Land use	Area (km^2)	%
Woodlands (including plantations)	644	30.2
Scrubland	283	13.2
Grassland and vlei	517	24.2
Cultivation and commercial farming	231	10.8
Cultivation and rural subsistence farming	261	12.2
Residential areas	146	6.8
CBD and avenues	5	0.2
Industrial areas	12	0.6
Hospitals	1	0.1
Lakes, dams, sewage farms	32	1.5
Other	4	0.2
Total	2136	100.0

residential suburbs is considerable, accounting for 6.8% of the catchment, or about one quarter of the total farming land available in the catchment. The City of Salisbury proper, including the Central Business District, the avenues, industrial areas, and hospitals, accounts for less than 1% of the total catchment area; the water requirements of this area are served by lakes and dams of twice that area (Fig. 6).

Population growth within the catchment

The population distribution maps (Figures 7 and 8) were prepared using estimates of the population of the catchment area derived from a 1969 census of the Seke Tribal Trust Land, and the cartographic symbols representing population centres on 1:50,000 physiographic maps. Whilst this method is far from satisfactory as the census figures and the physiographic maps used were out of date, and as the cartographic symbols were probably not entirely representative of the population distribution, the method was the best method of population estimation available.

The population distribution in 1951 was fairly sparse, especially in the Seke Tribal Trust Land, and the population of Salisbury was also small (Fig. 7). The figure of 119,000 persons for Salisbury is unrealistic for it includes all of the non-Africans but only the employed African population. On the other hand, Fig. 8, which illustrates the 1979 population distribution, appears to be inflated, although it is most probably a more realistic representation of the

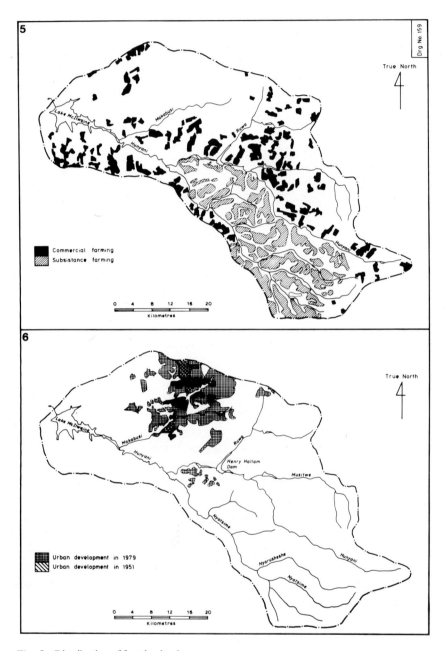

Fig. 5 Distribution of farming lands.
Fig. 6 Distribution of urban development.

18

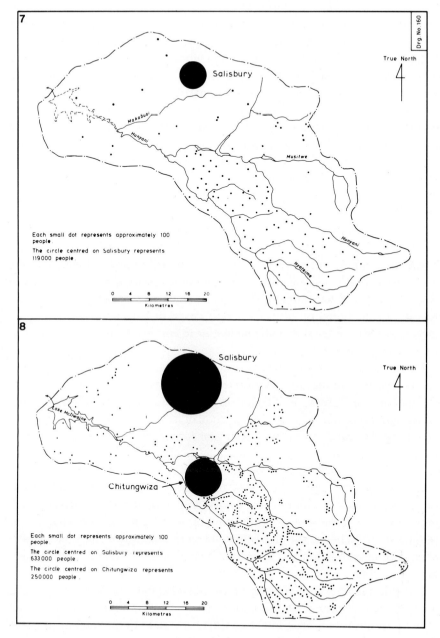

Fig. 7 Population distribution in the Lake McIlwaine area in 1951.
Fig. 8 Population distribution in 1979 showing urban expansion.

19

present population concentration in the catchment. Urban population has increased over eight-fold during the three decades shown in the figures, and the rural population has also increased considerably. A comparison of the population distribution pattern, Fig. 8, with the cultivation pattern, Fig. 5, indicates the even distribution of the population in the Seke Tribal Trust Land which is characterised by subsistence farming. The commercial farming area has a very low population density in comparison.

This expansion of both the urban and rural populations has intensified the pollution problems experienced in the Lake McIlwaine catchment and has led to greatly increased demands for potable water (see J. McKendrick, this volume; Mazvimavi, 1979). This pattern of population growth must be considered in the management of Salisbury's water resources.

Future development, of the Hunyani River catchment

There are a number of possible alternatives in the future development of the catchment area. These are given in the City of Salisbury's 'Strategic Planning Report' (City of Salisbury, 1976), and include two broad possibilities in relation to the pollution of water supply reservoirs: namely, to expand the urban area into other river catchments, or to confine the development to its present Hunyani River catchment.

Planners and other authorities concerned in the Hunyani River catchment should avoid polluting any further river catchments. For example, the City of Bulawayo, Zimbabwe, is situated in the Khami-Gwaai catchment, from which it draws but a very small proportion of its total water supply. The bulk of its supply comes from the Upper and Lower Ncema, Mzingwane, and Inyankuni Dams in the Upper Mzingwane catchment some 60 km to the east (Khupe, 1979), which is free of the influence of the City. In terms of Salisbury, a water supply strategy similar to that of Bulawayo would be the development of new dams in, for example, the Mazoe Valley on the Mazoe River catchment. Whilst the present water demand of the City might be inadequate to require such a scheme, the high water demand resulting from the rapid growth of the urban population and industrial expansion would make such an alternative feasible.

Considering the confinement of the development of Salisbury to the Hunyani River catchment, the City's development is based on a westward growth of the urban area into the Muzururu and Gwebi catchment areas (flowing into the Hunyani River at Lake Robertson), and this needs to be assessed. Another possibility would be to expand the industrial area between

Salisbury and Chitungwiza, thereby utilising the noisy area adjacent to the Salisbury Airport which is unsuitable for residential development. In that case, the expansion of Chitungwiza would be justifiable, and any potential pollution would be confined to the Hunyani River catchment upstream of Lake McIlwaine keeping Lake Robertson free of polluting loads.

Conclusions

1. The study reveals that the Upper Hunyani River catchment has an area of 2 136 km^2, 67% of which is covered by vegetation, 23% is cultivated by both commercial and subsistence farmers, and 10% is comprised of urban developments and the supporting water storage and treatment systems. It is believed that two thirds of the catchment is likely to maintain an ecological balance, whereas one third, consisting of the urban area and the cultivated rural area, is at risk and requires careful land management.
2. The land use in the Upper Hunyani River catchment has a direct bearing on water pollution. Due to the position of the urban area in the catchment, there is little doubt that the pollution of Lake McIlwaine is mainly the result of urban activities. The closeness of the high density residential areas to the main water supply reservoirs enhances the chances of pollutants being carried into these dams.
3. The location of the urban and industrial centres reduces the chances of the pollutants being broken down by natural processes. The problem stems from the fact that the City lies within the same catchment as its major water reservoirs and is likely to increase with the expansion of the urban areas, particularly the industrial and residential areas.
4. There is a need, therefore, to develop water supply reservoirs in areas where the possibility of water pollution is minimised such as in the Mazoe Valley, for example. However, with the Chitungwiza Urban Council set for further expansion and with no change of Government policy for the development of the City of Salisbury, the ecological instability of the Upper Hunyani River catchment is likely to be maintained for the foreseeable future.

Acknowledgements

This study was sponsored by the Institute of Mining Research, University of Zimbabwe, as part of a study of the geochemistry of this area. Its purpose was to provide data for the Hunyani River catchment as an aid in the interpretation of regional geochemical maps.

I would like to thank Messrs E. Mbwera, J. R. Whitlow, and R. G. Wheeler for their assistance, and I am most grateful to Professor K. A. Viewing for proposing and supporting this study, and for his encouragement.

References

City of Salisbury, 1976. Strategic planning report number 11: detailing and evaluation of the short list strategies. Unpublished report, Salisbury.

Kay, G. and M. Smout, 1977. Salisbury – a geographical survey of the capital of Rhodesia. Hodder and Stoughton, London.

Khupe, K., 1979. Changing patterns of water demand and supply in Bulawayo. B.A.(Hons.) thesis, University of Zimbabwe.

Mazvimavi, D., 1979. A survey of water resources and water pollution in Salisbury. B.Sc. (Hons.) thesis, University of Zimbabwe.

3 Physics

Physical limnology
P. R. B. Ward

Lake McIlwaine's physical limnology is controlled by the climate. The primary factors determining the surface and internal movements of the water are air temperature and winds, with river inflows becoming significant during the rainy season (see B. R. Ballinger and J. A. Thornton, this volume). A plan view and sections of the lake are given in Fig. 1. An excellent recent summary of physical mixing processes in lakes is given by Imberger (in Fischer *et al.*, 1979).

Although the underlying physical processes are the same, Lake McIlwaine differs from lakes in other parts of the world in seasonal timing. The climate in Zimbabwe is not typically tropical because it has a very definite cool season at one time of the year, and is not like northern temperate or Mediterranean climates because the rainy season is in summer in Zimbabwe rather than winter.

Maximum air temperatures are in October and November (Fig. 2) and maximum wind strengths are in September and early October (Fig. 3; Norton E.S.C. station, Department of Meteorological Services, unpublished data). The mean wind strength (2.6 m s^{-1}) is low compared with most parts of the world as is the maximum wind strength (9.0 m s^{-1} less than 1% of the time; Department of Meteorological Services, 1974). The prevailing wind direction at Lake McIlwaine is from the east (Fig. 3).

River inflows are moderate or large during the period from January to March and negligible during the months May to November (see B. R. Ballinger and J. A. Thornton, this volume). The coefficient of variation of the annual series of run-off of rivers supplying the lake (chiefly the Hunyani River) is large (80%) and hence the difference in inflows between a wet year and a dry year is very large. Years when the run-off volumes are as much as three times the long-term mean volume are not uncommon. Careful measurements of temperatures in both the lake and influent rivers (Thornton and

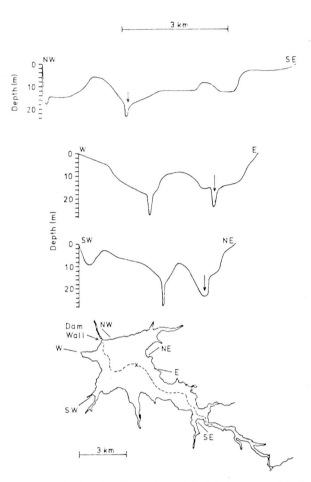

Fig. 1 Map of Lake McIlwaine showing sections of the lake bottom and the former river bed (dashed line). Mid-lake sampling station shown as an ×.

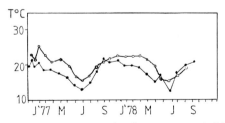

Fig. 2 Air (●) and water (o) temperatures during the two-year period 1976–78 (after Thornton, 1980).

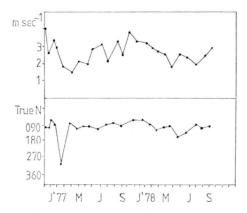

Fig. 3 Wind speed and direction at the Norton Electricity Supply Commision meteorological station during 1976–78.

Ward, 1978, unpublished data) show that entering river waters during major floods do not plunge to the bottom of the lake, but plunge to a depth of only 5 to 10 m below the surface.

Water balance

Over the long term there is obviously a balance between water entering the lake and water leaving it. Water volumes entering from the three major rivers (the Hunyani, Makabusi, and Marimba Rivers) and from local sources are equal to the volumes lost. These losses occur by water abstraction for City supply (via the Morton Jaffray Water Works), by evaporation from the lake surface and by downstream releases, either controlled releases or flood flow discharges over the spillway. Groundwater additions and losses are assumed to be negligible compared with the other amounts.

The exact amount of water abstracted by the City from year to year varies (see J. McKendrick, this volume), as does the amount of water passed over the spillway which during a wet year is large and during a dry cycle may be zero (see B. R. Ballinger and J. A. Thornton, this volume, for examples). During a long period of time, assuming the City is abstracting water at the maximum 'safe' yield, the amounts of water abstracted, evaporated, and passed over the spillway tend to be fixed percentages of the total water lost. These percentages are functions of reservoir volume and area, variability of annual run-off volumes from the contributing rivers, and other factors.

Values of these ratios are given in Mitchell (1974) in a generalised chart for reservoir yield estimation. The paper gives a numerical example for Lake McIlwaine without naming that lake. Table 1 shows these results. The calculation assumes a lake volume of 274×10^6 m^3, a coefficient of variation of the annual run-off series for the contributing rivers of 80%, and a net annual evaporation of 1700 mm. Mean annual volume entering the lake is determined by assuming a mean annual run-off of 120 mm over the 1700 km^2 river basin area. A curve of lake surface area versus water capacity based on known survey data for the Lake McIlwaine basin is used. A controlled abstraction rate equal to the 10% risk level (dry one year in ten) is assumed.

Table 1 Long-term average water balance; yearly volumes given

Inflow		Outflow		%
Source	Volume	Source	Volume	
Rivers	204×10^6 m^3	Controlled abstractions	123×10^6 m^3	60
		Evaporation	34×10^6 m^3	17
		Spillway discharges	47×10^6 m^3	23
Total	204×10^6 m^3	Total	204×10^6 m^3	100

Wind waves and seiches

No experimental programme of measuring either wind waves or seiches (long period, slow oscillations of the whole water body) has been undertaken. However, methods of calculation of both kinds of waves are reliable and well tested and may be used to make approximate predictions for Lake McIlwaine.

Wind waves. Under prevailing wind conditions, wave heights increase from east to west across the lake. After a time of about 1.5 hours (for Lake McIlwaine) from the commencement of the strong wind the waves become in equilibrium with the wind and a fully arisen sea is said to exist. Under this condition the significant wave height and significant wave period at the downwind shore of the lake may be predicted by the SMB (Sverdrup-Munk-Bretschneider) method (US Army Corps of Engineers, 1973). With a fetch of 5 km and a maximum sustained wind strength of 9 m s^{-1}, the SMB method gives the following results: a significant wave period of 2.7 s, and a significant wave height of 0.45 m. A wave of period 2.7 s would have a wave length of 11 m under deep water conditions. These approximate predictions are in agreement with wave conditions observed at the lake on very windy days.

Seiches and thermocline oscillations. Seiches are low frequency oscillations of the whole water body. During a fundamental mode (first harmonic) oscillation, the water surface slowly rocks back and forth, moving from end to end or from side to side of the lake. The existence of seiches have been documented for most major lakes in the world including Lake Kariba, Zimbabwe-Zambia (Ward, 1978; Ward, 1979). Seiches are driven by wind action on the lake surface. Strong winds blowing for a period of time about equal to half of the period of the first harmonic oscillation, and then reversing and blowing from the opposite direction, are particularly effective in causing seiches. The amplitude of seiche oscillations is usually small unless winds reach hurricane strength. No equipment for measuring seiches has been operated at Lake McIlwaine, and the wind waves are so much larger than the seiche oscillation that these would remain undetected.

The period of the first harmonic longitudinal seiche for Lake McIlwaine may be predicted approximately by determining the mean velocity (c) of a shallow water wave running the length of the lake. The shallow water wave travels at velocity $(g\,d)^{1/2}$, where g is the gravitational acceleration and d is the local depth.

Values of local depth may be taken from a longitudinal section of the lake (Fig. 1) and the mean value of the shallow water wave velocity determined. This is about 9.6 m s^{-1}. The period of the first harmonic seiche is:

$$T_1 = 2\,L\,/\,c \tag{1}$$

in which L is the length of the main part of the lake.

With L equal to 6000 m, the period of the first harmonic oscillation T_1 is 1250 seconds or approximately 20 minutes. The seiche may be expected to be most easily detectable in the narrow south-eastern part of the lake, where amplification of the oscillation occurs.

Although the water surface oscillation has little physical importance because it is so small, it will trigger a slow oscillation of much larger magnitude of the thermocline. This oscillation has important consequences for the hypolimnion: a series of movements of smaller magnitude are started and these movements control the mixing process in deep water.

The velocity of movement c_f of a disturbance at the interface of a two-layer fluid is given by Dyer (1972) as:

$$c_f = (\frac{\Delta \rho}{\rho}\,g\,d_s)^{1/2} \tag{2}$$

in which $\Delta \rho$ is the density difference (assumed to be small) between the two layers, ρ is the mean density of the two layers, and d_s is the depth of the upper layer.

Assuming that the quantities $(d_s)^{1/2}$ and $(d)^{1/2}$, which is the square root of the total depth, do not differ significantly from one another, the velocity of the interfacial wave is:

$$c_f = (\frac{\Delta \rho}{\rho} g \, d)^{1/2} \tag{3}$$

The period of oscillation T_{11} of the interface (inversely proportional to the wave velocity) will thus be:

$$\frac{T_{11}}{T_1} = \frac{(g \, d)^{1/2}}{(g \, d \, \Delta \rho / \rho)^{1/2}}$$

and

$$T_{11} = \frac{T_1}{(\Delta \rho / \rho)^{1/2}} \tag{4}$$

Lake McIlwaine is temperature stratified for several months of the year (see J. A. Thornton and W. K. Nduku, this volume), and a typical temperature difference across the thermocline is 2°C. This thermocline is weakly defined in the lake, but we will assume it exists for the purposes of this calculation. For the temperature range of interest in this study, an approximate water density versus temperature relationship is:

$$\Delta \rho / \rho = 0.000195 \, (\Delta \theta) \tag{5}$$

in which $\Delta \theta$ is the temperature difference (in °C). Thus, $\Delta \rho / \rho$ is equal to 0.00039.

Thus the period of the interfacial oscillation T_{11} is 17.5 hours. Strong winds which blow for about 9 hours down the axis of the lake and then stop or reverse will cause major oscillations of the thermocline.

The change in the horizontal level of the thermocline follows hydrostatic behaviour. Thus a surface perturbation is expected to be magnified ($\rho / \Delta \rho$) times the perturbation of the thermocline's position. With ($\rho / \Delta \rho$) equal to approximately 2500, it is clear that small surface perturbations show up very readily in thermocline behaviour. For example, an elevation of only 2 mm in the water surface level at one end of the lake would lead to a depression of 5 m in the thermocline. Table 2 summarises the information on water body oscillations. Note that these example magnitudes are not measured quantities but simply estimates of anticipated sizes.

Table 2 Calculated water body oscillations

Parameters	First mode longitudinal oscillation		
	Wave velocity m/s	Period hours	Example magnitudes of vertical displacement
Water surface	9.6	0.35	0.002 m
Thermocline	0.190	17.50	5.0 m

No measurements with continuous recording equipment measuring temperatures at many values of depth have been made to date, so it is not known whether these large oscillations in the thermocline's position occur.

Heat dynamics

Following a period of overturn during the winter, the lake commences its seasonal cycle in August (Fig. 4). Solar radiation and increasing air temperatures heat the surface waters causing the lake to become weakly stratified.

This stratification intensifies in September. Increasing wind strengths in September and early October (the windiest time of the year) cause vertical mixing of warm surface water and cooler water from the lower layers. This mixing reduces the vertical temperature gradient on windy days and is sufficient to cause a complete turnover only under exceptional conditions (see below: Potential energy of stratification). River inflows at this time of the year are negligible and do not play a role in the heat dynamics.

Water temperatures continue rising until about the middle of February. The heat dynamics during the period from December to February vary from season to season. During drier than average rainy seasons, the lake remains stably stratified during these months. In wetter than average seasons, low air temperatures for several days in succession combined with shear and turbulence from inflowing flood waters causes mixing in all but the bottom waters.

Air temperatures drop rapidly about the middle of April, and cause the lake to turn over at this time.

Temperature structure. The vertical variation of temperature at the centrally located station (shown as × in Fig. 1) was measured by R. D. Robarts during one year (1975–1976) when the rainy season was drier than average (Robarts and Ward, 1978). Significant river discharge did not occur until January

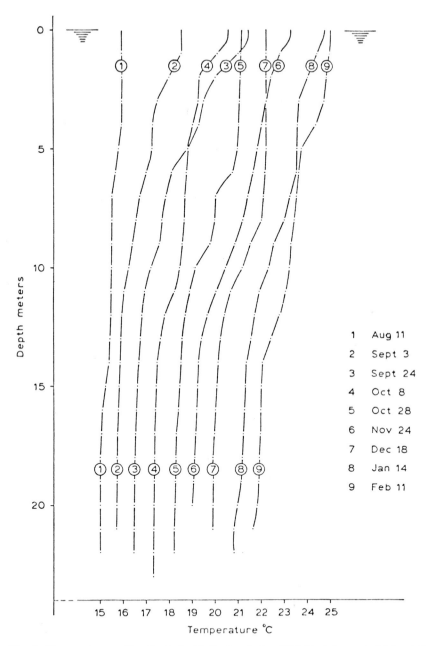

Fig. 4 Temperature profiles for Lake McIlwaine during 1975–76 at the central lake station shown in Fig. 1.

during this year. The maximum difference in temperature documented between the surface and bottom waters was 5°C (Fig. 4), on 24 September 1975. This stratified temperature profile was almost completely mixed by strong winds on 4 and 5 October. During the period from September to February, the difference in temperature between the surface and bottom waters was typically 3°C.

Heat transport in the hypolimnion. The energetics of mixing the hypolimnia of lakes is not well understood at present. Imberger and Patterson (1980; their Fig. 3.2) give a vertical profile of a lake, showing some of the mixing mechanisms. It is generally believed that wind-induced seiching leads to either wave or shear effects on the lake boundaries, or to wave effects within the water body. These cause a local overturning event which is followed by a readjustment of heat throughout the whole lake. The net effect of these processes is to transport heat downwards in the hypolimnion and to progressively raise its temperature over a long period. Although the process is not a simple diffusive transport-from-turbulence process, many workers (Jassby and Powell, 1975; Robarts and Ward, 1978; Imberger and Patterson, 1980) have used a simple one-dimensional diffusion equation to describe the vertical heat transport $\dot{M}$. Fick's equation for diffusion is:

$$\dot{M} = \bar{c} \rho K_z A \frac{\partial \theta_z}{\partial z} \tag{6}$$

in which $\bar{c}$ is the specific heat of water, ρ is the density of water, K_z is the vertical diffusion coefficient, A is the area of the horizontal slice considered, θ_z is the temperature, and z is the vertical co-ordinate.

By considering an elemental volume in the hypolimnion, and by equating the net heat input (derived from the derivative of vertical temperature gradient) with the increase of heat of the element over a period of time, values of K_z were derived (Robarts and Ward, 1978). The mean value of K_z over the range 12 to 18 m was 25 m^2 s^{-1}. This value is in accord with a trend for lakes from other parts of the world of similar surface area (Ward, 1977). The value is four orders of magnitude larger than the molecular diffusion coefficient for water at these temperatures.

Potential energy of stratification. When a body of water is stratified with slightly lighter than average water near the surface and slightly heavier than average water at the bottom, the centre of mass is lower than the centre of mass for homogeneous conditions. Potential energy is required to change the water from the stratified condition to the homogeneous condition. This energy is supplied by the wind (applied at the surface) or by inflowing river currents. During the period from August to December at Lake McIlwaine, river flows are negligible and hence all the energy for mixing is supplied by the wind.

The temperature profiles (Fig. 4) show that the heaviest stratification occurred on 24 September. Note that this stratification is weak compared with lakes in northern climates which undergo annual freeze cycles. The potential energy V required to change the profile from this stratified condition to the homogeneous condition is:

$$V = M g (\bar{z}_0 - z) \tag{7}$$

in which g is the gravitational acceleration, M is the mass of the water body, $\bar{z}_0$ is the centre of volume of the lake, and z is the centre of mass of the stratified lake. These centres are measured above any convenient datum, e.g., the bottom of the lake.

The centre of volume is the first moment of the area versus height curve divided by the total volume. Integrals are taken from the bottom ($z = 0$) to the surface ($z = H$):

$$\bar{z}_0 = \int_0^H A z \, dz \bigg/ \int_0^H A \, dz \tag{8}$$

The centre of mass is the first moment of the mass versus height curve divided by the total mass:

$$z = \int_0^H \rho A z \, dz \bigg/ \int_0^H \rho A \, dz \tag{9}$$

With the values of A from Robarts and Ward (1978) and values of ρ from the temperature profile for 24 September and equation (5), and integrating from 20 m below the surface ($z = 0$) to the surface ($z = 20$ m), the difference between the centres of volume and mass is: $\bar{z}_0 - z = 12.8255 - 12.8243 = 0.0012$ m. The potential energy of stratification V, from equation (7), with M equal to 274×10^9 kg, is thus 3.22×10^9 newtons. On other days (Fig. 4) the potential energy of stratification will be smaller, and occasionally it may be larger than this value.

The time T_M required to cause the lake to change from a stratified to a homogeneous density distribution is the quotient of the potential energy and the power available for overturning:

$$T_M = V / P \tag{10}$$

The power available for overturning is supplied by the wind, and is very much less than the rate at which it is working at the lake surface. This is because parts of the lake surface are sheltered from the wind and because a large fraction of the energy is lost by turbulence in the upper layers of the water body and dissipated as very small increments of heat energy. The power from the wind available for mixing of the lower water column is given by Imberger, et al. (1978) as:

$$P = \tilde{m} \, C_D \, \rho_a \, A \, U^3 \tag{11}$$

in which $\tilde{m}$ is a constant equal to 0.0012, C_D is the drag coefficient for wind-water shear stress (equal to 0.0013), ρ_a is the density of air, A is the area of the lake surface, and U is the wind velocity at 10 m.

With ρ_a equal to 1.25 kg m^{-3} and A equal to 26.3 × 10^6 m^2, the power available P may be computed and hence the turnover time T_M for various values of the wind velocity U. The cube law relationship given in equation (11) means that these values are very sensitive to the assumed values of wind velocity. Table 3 gives the results. The strongest sustained winds normally blow during daylight hours. Two successive days of winds of speed 9 m s^{-1} would thus be required to overturn the lake. Winds of this magnitude blowing for several hours continuously are of rare occurrence but are possible during the windy period from September to mid-October. Years when overturn of the lake during this period occurs must therefore be expected (see J. A. Thornton and W. K. Nduku, this volume, for an example).

Table 3 Wind durations required for overturning a stratified lake*

Wind velocity at 10 m m s^{-1}	Power available for overturning watts	Time required days
12	88,600	0.36
9	37,400	0.86
5	6,410	5
3	1,390	23

* Assuming a 5°C temperature differential between surface and bottom waters (e.g., 24 September temperature profile; Fig. 4).

References

Department of Meteorological Services, 1974. Climate information sheet No. 54, Maximum wind values. Salisbury.

Fischer, H. B. *et al.*, 1979. Mixing in inland and coastal waters. Academic Press, New York.
Dyer, K. R., 1972. Estuaries: a physical introduction. Wiley, London.
Imberger, J., 1979. Mixing in reservoirs. In: H. B. Fischer *et al.*, Mixing in inland and coastal waters. Academic Press, New York.
Imberger, J. and J. C. Patterson, 1980. A dynamic reservoir simulation model – Dyresm 5. Paper presented at Symposium on Predictive Abilities of Surface Water Flow and Transport Models. University of California, Berkeley, Berkeley.
Imberger, J., J. C. Patterson, B. Hebbert and I. Loh, 1978. Dynamics of a reservoir of medium size. J. Hydraul. Div. Am. Soc. civ. Engrs., 104 (HY5): 725–743.
Jassby, A. and T. Powell, 1975. Vertical patterns of eddy diffusion during stratifications in Castle Lake, California. Limnol. Oceanogr., 20: 530–543.
Robarts, R. D. and P. R. B. Ward, 1978. Vertical diffusion and nutrient transport in a tropical lake (Lake McIlwaine, Rhodesia). Hydrobiol., 59: 213–220.
Thornton, J. A., 1980. Factors influencing the distribution of reactive phosphorus in Lake McIlwaine, Zimbabwe. D.Phil. Diss., University of Zimbabwe.
US Army Corps of Engineers, 1973. Shore protection manual. Volume I. Coastal Engineering Research Center, Fort Belvoir, Virginia. US Government Printing Office, Washington.
Ward, P. R. B., 1977. Diffusion in lake hypolimnia. I.A.H.R. Proceedings of the Baden Baden Symposium. Paper No. A88, pp. 103–110.
Ward, P. R. B., 1978. Water surface fluctuations at Lake Kariba. The Rhod. Engr., Paper No. 195, 16: 133–142.
Ward, P. R. B., 1979. Seiches, tides, and wind set-up at Lake Kariba. Limnol. Oceanogr., 24: 151–157.

The hydrology of the Lake McIlwaine catchment
B. R. Ballinger and J. A. Thornton

Hydrology is the study of water and in particular its movement over, under and above the ground. Thus, hydrology includes aspects of meteorology such as precipitation and evaporation, stream flow, transpiration, and underground water; hydrology may even include the abstraction of water. In Zimbabwe, extensive monitoring of most of these hydrological parameters is carried out. Meteorological parameters (rainfall and evaporation) are measured by the Department of Meteorological Services which maintains an extensive network of stations around the country. Similarly, the Hydrological Branch of the Division of Water Development maintains a network of river and groundwater flow monitoring stations throughout the country.

In relation to Lake McIlwaine, the Department of Meteorological Services regularly monitors rainfall and evaporation at 17 stations in the Upper Hunyani River catchment; one station, which will be referred to in this section, is located on the southern shore of Lake McIlwaine near the National Park office. River flow is monitored at 18 stations within the lake catchment

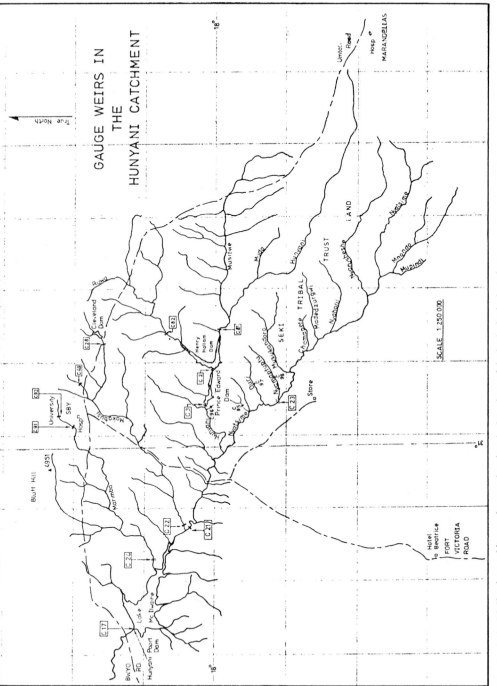

Fig. 5 Lake McIlwaine and its catchment showing the locations of the hydrological stations.

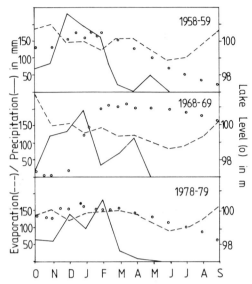

Fig. 6 Mean monthly evaporation and precipitation in mm at Lake McIlwaine during 1958–59, 1968–69 and 1978–79. Changes in lake level (o) are also shown.

by the Hydrological Branch. Four gauging stations are of particular relevance (Fig. 5): namely, C21, C22, and C24 on the inflowing Hunyani, Makabusi, and Marimba Rivers, and C17 on the Hunyani River outflow. Rainfall and evaporation records date back to 1957–58, whilst flow records begin in 1953–54 (except in the case of C21 which began recording in 1957–58). Both meteorological and river flow data are published annually in the form of data summaries, but little other published data are available.

Meteorological conditions

Precipitation in the Lake McIlwaine catchment, as in all of Zimbabwe, is highly seasonal with quite distinct wet and dry seasons (Fig. 6). Three seasons are readily discernable (Vincent *et al.*, 1960; Marshall and Falconer, 1973; Thornton, 1980a). 'Spring' is a hot dry season which falls between September and November. Intense rainfall is unlikely to occur although light falls may be expected. Average daily temperatures are approximately 22°C ± 6°C. 'Summer' is commonly known as the rainy season, and is characterised by being hot and wet. This season covers the period between December and April, and average daily tempeatures are approximately 20°C ± 6°C. The third season is cold and dry, and corresponds to 'Winter' in other parts of the world. Winter occurs between May and August when temperatures average 14°C ± 7°C.

Mean total annual rainfall is approximately 704 mm at Lake McIlwaine, but ranges considerably between years from a recorded low of 410 mm to a high of 1236 mm (Fig. 7). However, the quasi-20-year oscillation in rainfall observed in the summer rainfall areas of South Africa by Tyson (1978) and Allanson (1979) is not clearly defined in the Lake McIlwaine data. On the other hand, certain trends in their data are reflected at Lake McIlwaine. The year 1967 was a below average rainfall year in a decade where rainfall was largely below normal, and 1976 was centred in a decade where precipitation was above average as perhaps was 1958 (Allanson, 1979). Alternatively, an approximately five-yearly cycle has been suggested for Zimbabwe (D. S. Mitchell, personal communication; Thornton, 1980b). Given this latter period of oscillation, years centred in periods of above average precipitation are 1962, 1968, 1973, and 1976 whilst 1963, 1967, 1972, and 1975 marked periods of below average rainfall. This periodicity in rainfall is reflected in lake level variations (Fig. 7).

Evaporation was generally inversely related to rainfall and was invariably higher. Total annual evaporation (as measured in class A evaporation pans, painted and screened type) ranged from 1291 mm to 2005 mm, and averaged 1541 mm. Evaporation rate generally decreased over the twenty-year period of record (Fig. 7). However, it is not possible to link these variations with perturbations in the micro-climate of the Lake McIlwaine area, despite the seeming desirability of linking decreased evaporation to the eradication of water hyacinth from the lake surface and the cessation of evaporative losses through transpiration.

Surface run-off

River flow like rainfall is highly seasonal. Main river flows normally occur between December and April, although residual flows resulting from run-off from the sewage farms and the release of compensation water from the upstream dams do occur at other times (Fig. 8). Riverine inflows usually exceed outflows although this situation may be reversed toward the end of the season when inflows taper off (as in 1958–59; Fig. 8). Mean annual run-off averages 304.7×10^6 m^3, but total annual run-off can range from 20.7×10^6 m^3 to 796.5×10^6 m^3 during below and above average rainfall years respectively.

River flow is the principal component of the water budget of the lake (see P. R. B. Ward, this volume, and below), with the Hunyani River contributing

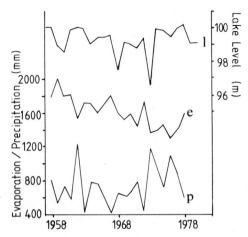

Fig. 7 Total annual evaporation (e) and precipitation (p) in mm at Lake McIlwaine between 1958 and 1978, and mean annual lake level (l) variations over the same period.

upwards of 80% of the total gauged inflow, with the Makabusi and Marimba Rivers accounting for the remainder. Ungauged flows amount to about 25% of the total gauged flow and approximately 15% of the total annual inflow volume. Abstraction and downstream release account for the bulk of the outflow from the lake, about 60%, whilst evaporative losses amount to approximately 30% of the outflow. Since the construction of the downstream Darwendale Dam (Lake Robertson), the amount of water released from Lake McIlwaine has been somewhat reduced but this has had little effect on the overall percentage of water leaving the lake. Groundwater inflows and outflows are considerd to be minimal in comparison with the surface flows.

Figure 9 shows the year to year variation in inflow and outflow of Lake McIlwaine. These variations are closely related to rainfall and follow a similar pattern as comparison of Figs. 7 and 9 will show.

Lake level

Lake levels normally vary within a range of about two metres of full supply level (FSL) per annum largely in response to abstraction by the City of Salisbury for water supply purposes and to satisfy downstream demands (Fig. 6). Draw-downs of up to four metres can however occur during below average rainfall periods when there is reduced inflow. Lake levels follow a seasonal pattern that is related to rainfall and riverine inflow (Fig. 9). The very significant decline in lake level during 1968 had major ecological effects,

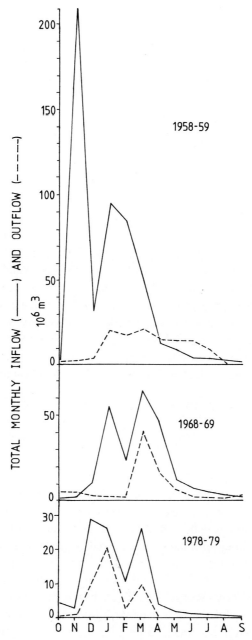

Fig. 8 Seasonal variation in riverine inflows and outflows during 1958–59, 1968–69 and 1978–79.

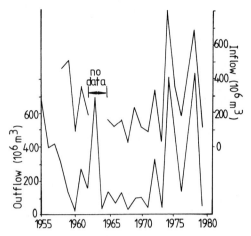

Fig. 9 Variation in total annual inflow and outflow of Lake McIlwaine between 1954 and 1979.

particularly on the water chemistry and benthic fauna of the lake (Marshall and Falconer, 1973; Marshall, 1978). Similar effects probably occurred again in 1973. Such drastic drops in water levels are unusual, but do appear to be related to the quasi-5-year cycle in precipitation noted above. These lake level variations often affect other lake users such as boat owners and marinas as well as the bottom fauna and water chemistry. Since the construction of the Darwendale Dam these variations in lake level have been greatly reduced as it is no longer necessary to release water from Lake McIlwaine for downstream users.

Water budget

Marshall and Falconer (1973) have presented a water budget for Lake McIlwaine for a number of years in the late 1960s. As these budgets reflect both above and below average years, they present a useful summary of the hydrological characteristics of the Lake McIlwaine system as well as a corollary to the long-term mean water balance given by P. R. B. Ward (this volume). Data for two years, 1966 and 1968, are presented in Table 4 (Marshall and Falconer, 1973; Table 4). Low riverine inflow during 1968 combined with below average rainfall resulted in the inflow deficit observed during that year, whilst 1966 was more normal. Such variations yield a wide range of water retention times, although in most instances the lake is flushed each year. The theoretical water retention time is 0.8 years and is calculated from the lake volume and the mean annual inflow volume (Marshall and

Table 4   Water budget for Lake McIlwaine for 1966 and 1968 (after Marshall and Falconer, 1973)

Parameter	Year	
	1966	1968
Inflow ($\times 10^6$ m^3)		
Total gauged flow	101.1	19.7
Estimated ungauged flow	24.7	4.9
Direct rainfall	18.5	9.9
Total	144.3	34.5
Outflow ($\times 10^6$ m^3)		
Abstractions	24.7	34.5
Downstream release	58.0	16.0
Evaporation	40.7	41.9
Total	123.4	92.4

Falconer, 1973; N. A. Burke and J. A. Thornton, this volume). This low water retention time greatly influences the chemical environment of the lake as well as the biological community as will be shown in the following sections.

References

Allanson, B. R., 1979. The physico-chemical limnology of Lake Sibaya. In: B. R. Allanson, ed., Lake Sibaya. Monographiae Biologicae, 36: 42–47.
Marshall, B. E., 1978. Aspects of the ecology of benthic fauna in Lake McIlwaine, Rhodesia. Freshwat. Biol., 8: 241–249.
Marshall, B. E. and A. C. Falconer, 1973. Physico-chemical aspects of Lake McIlwaine (Rhodesia), a eutrophic tropical impoundment. Hydrobiol., 42: 45–62.
Thornton, J. A., 1980a. Factors influencing the distribution of reactive phosphorus in Lake McIlwaine, Zimbabwe. D.Phil. Diss., University of Zimbabwe.
Thornton, J. A., 1980b. A review of limnology in Zimbabwe: 1959–1979. NWQS Rep. No. 1, Ministry of Water Development and Department of National Parks and Wild Life Management, Causeway, Zimbabwe.
Tyson, P. D., 1978. Rainfall changes over South Africa during the period of meteorological record. In: M. S. A. Werger, ed., Biogeography and ecology of southern Africa. Junk, The Hague.
Vincent, V., R. G. Thomas and R. R. Staples, 1960. An agricultural survey of southern Rhodesia. Part I: Agro-ecological survey. Government Printer, Salisbury.

4 Chemistry

Water chemistry and nutrient budgets
J. A. Thornton and W. K. Nduku

The study of the chemical limnology of Lake McIlwaine spans some twenty years, and although the chemical records are not continuous throughout this period it is nevertheless reasonable to say that the chemical limnology of the lake is well known. The first studies were conducted in the early 1960s as the lake entered a mesotrophic phase and subsequent studies have traced the chemical environment of the lake progressively through eutrophy and hypereutrophy. Following the diversion of municipal wastewater to pasture irrigation schemes in the early 1970s, most recent studies have followed the recovery of the lake to near mesotrophy. This section describes the chemical environment of the lake and outlines in physico-chemical terms the eutrophication and recovery of Lake McIlwaine.

Water chemistry

Thermal and oxygen regimes. Lake McIlwaine has been described as a warm (tropical) monomictic lake by numerous researchers (Van der Lingen, 1960; Munro, 1966; Marshall and Falconer, 1973a, 1973b; Mitchell and Marshall, 1974), although more recent studies have shown that the lake may be polymictic at times given proper conditions of surface cooling and riverine inflow (Nduku, 1978; Thornton, 1980a; P. R. B. Ward, this volume). Van der Lingen (1960) identified a strong, classically-defined thermocline in the lake during the summer of 1957–58. He also noted the presence of a surface to bottom oxygen gradient with de-oxygenation existing below 10 m for part of the year. Similar trends were described by Munro (1966) during 1961–62. Subsequent workers have failed to find so distinct a thermocline, although a strong oxycline has been observed, usually at about the 10 m level (Marshall and Falconer, 1973a; Mitchell and Marshall, 1974; Nduku, 1978; Thornton,

1979a, 1980a, 1980b).

Figure 1 shows the thermal and oxygen regimes in Lake McIlwaine during 1969 and 1979 at the mid-lake station, SM-4 (Fig. 2). The lake was monomictic during both years with stratification occurring in September-October and overturn in February-March. As noted, the thermocline is poorly defined although the oxycline is quite distinct during the periods of stratification. Figure 3 shows a similar pattern for 1977, although the lake was polymictic during the period between January and April of that year. This polymictic condition is seen most clearly in the oxygen saturation curve which shows re-oxygenation of the hypolimnion on two occasions during January and February (1977) and a fluctuating hypolimnion during March. Overturn took place during April. Similar fluctuations of the oxy-thermocline were observed by Nduku (1978) during 1975–76. On both occasions, nutrient up-

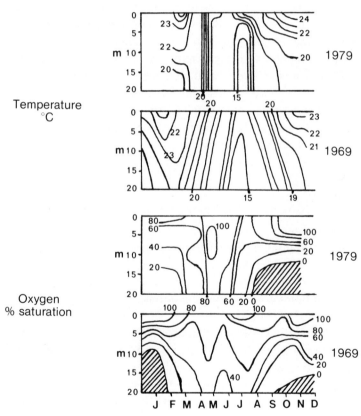

Fig. 1 Annual temperature and oxygen cycles in Lake McIlwaine during 1969 and 1979; data for 1969 re-drawn from Marshall and Falconer (1973a).

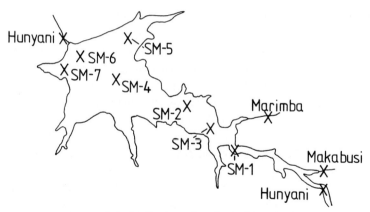

Fig. 2 Station locations on Lake McIlwaine referred to in this and subsequent papers. Station SM-4 is the mid-lake station.

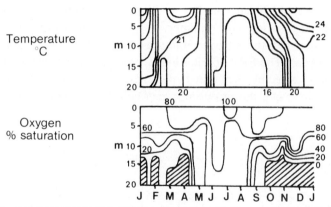

Fig. 3 Annual temperature and oxygen cycle in Lake McIlwaine showing polymicism during 1977.

welling and temporarily localised algal blooms confirmed the fact that mixing had taken place (Nduku, 1978; Thornton, 1980a).

The annual temperature range observed in Lake McIlwaine is from a minimum of 14°C in mid-winter (July) to a maximum of usually no higher than 25°C, although extreme surface temperatures of up to 28°C have been recorded in mid-summer (January). Surface to bottom temperature gradients range up to a maximum difference of 6°C over the water column. Oxygen saturations range from nil to over 100%, with values of 125% being not uncommon in the surface waters of the lake during summer and spring algal blooms.

Dissolved nutrients. Nitrate and ammonia nitrogen and soluble reactive phosphorus (SRP) are amongst the most important forms of the plant nutrients, nitrogen and phosphorus, found in Lake McIlwaine (Marshall and Falconer, 1973a); nitrite nitrogen and particulate phosphorus are of lesser importance and vary only slightly during the year (Nduku, unpublished data; Thornton, 1980a). Nitrite and particulate phosphorus concentrations at the mid-lake station are usually less than 0.005 mg l^{-1}, whilst concentrations of nitrate, ammonia and SRP are generally in excess of 0.010 mg l^{-1}.

The seasonal distributions of the nitrate and ammonia nitrogen fractions are shown in Figs. 4 and 5 respectively for surface waters at the mid-lake station during 1969 and 1979. Similar distributions are seen for both years with nitrate being inversely related to ammonia concentrations. Generally, nitrate maxima occur in spring and summer with minimum values occurring during winter, coincident with increases in chlorophyll *a* (Marshall and Falconer, 1973a; Thornton, 1980a). On the other hand, maximum ammonia values are observed during the late winter months with minima generally occurring during spring and summer, suggesting an oxidation-reduction relationship between the nitrogen fractions (Marshall and Falconer, 1973a; Thornton, 1980a). The SRP distribution is similar to that of nitrate (Fig. 6), and is also influenced by algal growth patterns, during both 1969 and 1979.

Whilst, on average, the nitrate nitrogen concentrations observed during

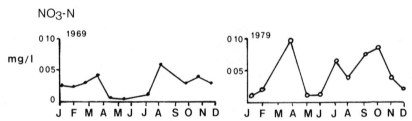

Fig. 4 Distribution of NO_3-N in the surface waters of the lake at SM-4 during 1969 and 1979.

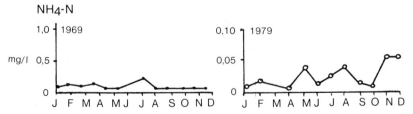

Fig. 5 Distribution of NH_4-N in the surface waters of the lake at SM-4 during 1969 and 1979.

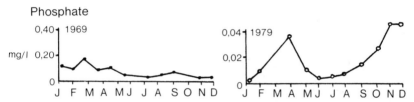

Fig. 6 Distribution of SRP in the surface waters of Lake McIlwaine at the mid-lake station during 1969 and 1979.

1979 are slightly higher than those recorded during 1969, it is interesting to note that the ammonia and SRP concentrations are approximately an order of magnitude less during 1979 than during 1969. This is undoubtedly a function of the diversion of treated sewage effluent from the lake inflows, and will be discussed further in a later part of this section.

Inorganic ions. Lake McIlwaine is fairly typical of most southern African man-made lakes in terms of its inorganic chemistry (Walmsley and Toerien, 1977; Scott *et al.*, 1977; Grobbelaar *et al.*, 1980). Most inorganic ions follow similar seasonal trends, with maximum values being recorded in spring and summer, and minimum in winter. This pattern is closely related to the hydrological regime (see B. R. Ballinger and J. A. Thornton, this volume) and reflects concentration by evaporation during the hot spring and summer months as well as dilution by riverine inflows during late summer and winter. Figures 7, 8 and 9 show the seasonal distribution of conductivity, pH and alkalinity, respectively, in the surface waters at the mid-lake station during 1969 and 1979. Similar trends are seen during both years.

The range in conductivity is from a minimum of 6.5 mS m^{-1} reported by Thornton (1980a) to a maximum of 23.2 mS m^{-1} recorded by Marshall and Falconer (1973a). Thus, Lake McIlwaine may be classed as a soft water, Class I (less than 60 mS m^{-1}) African lake using the Talling and Talling (1965) classification system. The lake is also usually slightly alkaline with a pH range of between 6.3 and 9.8. Alkalinity ranged from 40 to 60 mg $CaCO_3$ l^{-1} in most studies.

The seasonal distributions of the major cations found in the surface waters of Lake McIlwaine are shown in Figs. 10, 11 and 12. Figures 10 and 11 show the distributions of iron and manganese respectively at the mid-lake station during 1969 and 1979. The seasonal variations of sodium, potassium, magnesium and calcium at the same station are shown in Fig. 12 for 1979 only.

Potassium, magnesium, calcium and manganese concentrations remain

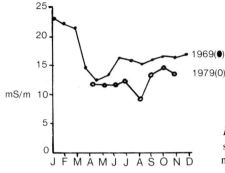

Fig. 7 Conductivity distribution in the surface waters of Lake McIlwaine at the mid-lake station during 1969 (●) and 1979 (o).

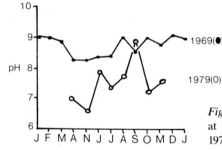

Fig. 8 pH distribution in the surface waters at the mid-lake station during 1969 (●) and 1979 (o).

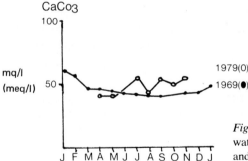

Fig. 9 Alkalinity distribution in the surface waters at the mid-lake station during 1969 (●) and 1979 (o).

relatively constant throughout the year, although there is a slight increase in concentration in spring probably due to concentration by evaporation. Iron and sodium show somewhat similar trends through the year and have slightly more variation in their annual concentration range. Sodium is the dominant cation, although data presented by Marshall and Falconer (1973b) suggest that calcium may have been dominant, or at least co-dominant, prior to the onset of eutrophication in the lake (circa 1960). The order of dominance of the

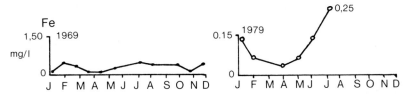

Fig. 10 Distribution of iron in the surface waters at station SM-4 during 1969 (●) and 1979 (o).

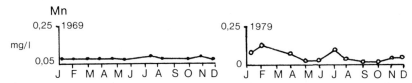

Fig. 11 Distribution of manganese in the surface waters at station SM-4 during 1969 (●) and 1979 (o).

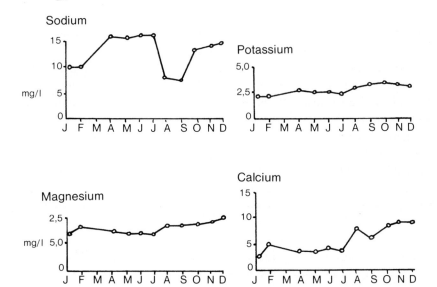

Fig. 12 Distribution of the major cations (sodium, potassium, magnesium and calcium) in the surface waters of Lake McIlwaine at station SM-4 during 1979.

major cations in Lake McIlwaine is Mn < Fe < K < Mg < Ca < Na, in the ratio of 1 : 2 : 33 : 60 : 85 : 190 during 1979. The concentration of iron has significantly decreased since 1969 (Fig. 10).

Little recent data are available on the distribution of anions in the surface waters of Lake McIlwaine. However, the data that are available suggest that there has been little change in the anionic distribution since the work of Marshall and Falconer (1973b). Chloride is the dominant anion followed by sulphate.

Nutrient budgets

The first estimate of riverine nutrient loadings to Lake McIlwaine was made by Marshall and Falconer (1973b) for SRP and for nitrate and ammonia nitrogen. More recently, Thornton (1979a, 1979b, 1980a) has estimated the nutrient budget of the lake using total reactive phosphorus (TRP) and ammonia and nitrate nitrogen; the former is approximately equivalent to 'total' phosphorus (Thornton, 1980a). The phosphorus and nitrogen loadings computed by Marshall and Falconer (1973b) and Thornton (1980a) are summarised in Table 1.

Due to the seasonal nature of the hydrological regime, nutrient loading to Lake McIlwaine is highly seasonal with upwards of 80% of the annual nutrient load entering the lake during the summer rains. Much of the nutrient load to the lake, however, is removed from the water column by abiotic means and is deposited in the lake sediments (Nduku, 1976; Thornton, 1979b, 1980a; J. A. Thornton and W. K. Nduku, this volume). Despite predictions of the occurrence of deleterious effects on the lake's water quality through internal nutrient loading (Robarts and Ward, 1978), there is little evidence that internal nutrient recycling is of great significance at the present time (Thornton, 1979b, 1980a). In fact, there has been a large reduction in the phosphorus load to the lake in the decade between the two studies (Table 1). An examination of the reasons for this reduction in phosphorus loading between 1969 and 1979 is made below. Little change in the nitrate nitrogen load has been observed, although the overall nitrogen load to the lake has also been substantially reduced due to the decline in the ammonia nitrogen load. This reduction is also discussed below.

Eutrophication and recovery

The assessment of the effects of the input of treated sewage effluent and its subsequent diversion from a lake can take two forms: namely, an examination of the changes in the various physico-chemical indices, and an examination of the changes in the level of nutrient loading. Both of these methods have been adopted in this section.

Table 1 Reactive phosphorus and ammonia and nitrate nitrogen loadings to Lake McIlwaine during 1967 (after Marshall and Falconer, 1973b) and 1977 (after Thornton, 1979a, 1980a) in metric tonnes P and N

Source	1967			1977		
	SRP	NO_3-N	NH_3-N	TRP	NO_3-N	NH_3-N
Inputs						
Hunyani	3.2	3.6	0.9	31.8	93.1	6.5
Makabusi	183.5	121.4	139.1	20.5	58.1	2.7
Marimba	101.4	33.3	17.8	29.3	35.0	1.1
Sediments	—	—	—	16.1	?	?
Totals	288.1	158.3	157.8	97.7	186.2	10.3
Outputs						
Hunyani				27.2	252.2	13.0
Water-works		no data		1.2	2.2	0.4
Algal uptake				3.4	?	?
Sedimentation				54.0	?	?
Totals				85.8	254.4	13.4
Change in solution in lake				−0.8	+10.9	−3.1
Mean amount in solution in lake				8.4	76.0	16.0
Loading rates						
Areal (g m^{-2})	11.0	6.3	6.3	3.9	7.5	0.4
Volume (g m^{-3})	1.2	0.6	0.6	0.4	0.8	0.1
Total inflow (10^6 m^3)		128.3			433.8	

Thirteen common trophic state indicators have been monitored both prior to and following the implementation of the nutrient diversion scheme. These indices are summarised in Table 2 for surface waters at the mid-lake station in Lake McIlwaine. The deterioration of the water quality of the lake between the earliest studies of 1957 to 1963 (Van der Lingen, 1960; Munro, 1966; Marshall and Falconer, 1973a, 1973b) and the middle period studies of 1968 to 1975 (Falconer *et al.*, 1970; Marshall and Falconer, 1973a, 1973b; Mitchell and Marshall, 1974; Robarts and Mitchell, 1976; Robarts and Southall, 1977; Robarts, 1979) can be clearly seen. An improvement in the water quality of the lake can be seen from 1976 onwards (Thornton, 1979a, 1980a; Hydrobiology Research Unit, University of Zimbabwe, unpublished data). Mean values of the indices measured show a return to the levels observed before the

Table 2 Comparison of the surface water chemistry of Lake McIlwaine at the mid-lake station between 1957 and 1980: data from various sources referred to in the text. Mean values are given immediately below the range of concentrations. Concentrations are in mg l^{-1} except where noted

Parameter	1957–58	1962–63	1968–69	1970–71	1974–75	1976–77	1977–78	1978–79	1979–80
Nutrients									
SRP	–	tr–0.13	tr–0.80	0.03–0.40	0.03–0.13	tr–0.30	tr–0.44	tr–0.04	tr–0.08
		0.04	0.16	0.22	0.08	0.04	0.04	0.01	0.04
NO_3-N	–	tr–0.15	tr–0.47	–	0.32–0.68	0.03–0.68	0.02–0.34	0.01–0.20	tr–0.20
		0.03	0.05	0.06	0.48	0.31	0.12	0.06	0.06
NH_4-N	–	tr–0.23	tr–2.0	–	0.03–0.11	tr–0.19	tr–0.08	tr–0.40	tr–0.14
		0.11	0.40	0.05	0.08	0.03	0.03	0.06	0.04
NO_2-N	–	tr–0.036	–	–	tr–0.008	tr–0.033	tr–0.007	tr–0.007	tr–0.024
		0.009			0.005	0.007	0.002	0.002	0.010
Inorganic ions									
Cond. mS m^{-1}	8.2–10.2	8.6–10.8	12.3–23.2	–	–	4.6–32.0	4.4–24.5	–	9.0–15.0
	8.8	9.5	17.2	16.4		1.4	10.1		12.6
pH	7.5–8.0	6.8–9.0	8.0–9.6	–	7.4–10.1	6.4–9.8	6.3–8.9	–	6.5–9.1
		7.5	8.6	9.0	8.3	7.2	7.2		7.5
Alkalinity (CaCO$_3$)	40	20–43	–	50	–	–	–	–	40–60
		38	50						45
Na	–	–	–	–	8.0–10.0	6.1–14.8	8.5–10.7	8.1–16.8	13.7–14.1
		14.0	23.5	23.5	9.0	10.6	9.5	12.7	13.9
Ca	–	5.3–11.3	–	–	5.9–10.6	2.4–10.2	1.9–3.2	2.5–8.1	8.2–8.6
		7.4	12.5	15.0	7.7	6.4	3.3	4.8	8.4
Mg	–	–	–	–	2.7–3.2	2.8–9.2	1.5–3.0	1.5–4.7	4.7–4.8
		9.0	12.5		3.0	7.4	2.6	3.7	4.8
K	–	–	–	–	1.7–2.7	2.2–4.8	2.4–2.6	2.3–3.3	3.1–3.3
				4.5	2.1	3.0	2.5	2.7	3.2
Cl	–	7.5	17.5	–	–	–	–	–	13–17
									15.0
SO$_4$	–	1.0–9.0	–	–	–	–	–	–	–
		3.4	10.0						
Chlorophyll a mg m^{-3}	–	–	50–150	–	12–95	tr–146	tr–142	4–14	2–29
						15.0	15.0	10.0	9.0
Secchi disc m	1.5–2.25	–	–	0.45–1.0	–	0.5–4.0	1.0–2.8	–	1.5–2.0
					1.2	1.3	1.2		1.8

lake became eutrophied, although the ranges in concentration remain fairly large.

Mean concentrations of SRP and ammonia in Lake McIlwaine have been reduced from 0.16 mg P l^{-1} and 0.40 mg N l^{-1} respectively in 1968–69 to 0.04 mg l^{-1} since 1976. This is a reduction of over 70% in terms of phosphorus concentration, and of 90% in terms of ammonia concentration (Figs. 5 and 6). Similar reductions in phosphorus concentrations have been observed in the Makabusi and Marimba Rivers. Marshall and Falconer (1973b) reported a range in SRP concentration of between 5 and 20 mg P l^{-1} in these rivers whilst Thornton (1979a, 1980a) recorded values of 0.1 to 1.7 mg P l^{-1}. Nitrate concentrations, unlike most other trophic state indicators in Table 2, have continued to increase following the diversion of wastewater nutrients to pasture irrigation schemes (Fig. 4). No improvements were noted until after 1978, and then the reduction in concentration was only to 1970–71 levels. Nitrite concentrations have remained largely unaffected. The apparent lack of effect of nutrient diversion on the nitrate and nitrite fractions is not entirely unexpected as nitrogen is virtually unaffected by pasture irrigation, and may even be enhanced by some pasture crops (see J. McKendrick, this volume).

Coincident with the overall reduction in phosphorus and nitrogen concentrations, the latter primarily through the reduced ammonia concentrations, has been a shift in phytoplankton growth limiting nutrient from the potential nitrogen limitation observed by Robarts and Southall (1977) to a joint phosphorus-nitrogen limitation (Watts, 1980; C. J. Watts, this volume; Thornton, 1980a). This shift, however, has not seemingly affected the phytoplankton standing crop as determined by chlorophyll *a* analysis; chlorophyll *a* concentrations of up to 146 mg m^{-3} have been observed during bloom conditions as recently as 1977 (Thornton, 1979a, 1980a). Nevertheless, mean phytopigment concentrations recorded since 1976 would seem to represent a much reduced level of phytoplankton standing crop when compared to values given by Falconer *et al.* (1970) and Robarts (1979). Certainly, the mean annual chlorophyll *a* concentrations of 9 to 15 mg m^{-3} recorded between 1976 and 1980 are well below the minimum recorded chlorophyll *a* concentration of 50 mg m^{-3} reported during 1968–69 (Falconer *et al.*, 1970), and compare favourably with the minimum value of 12 mg m^{-3} reported by Robarts (1979) in 1974–75. This reduced standing crop has not reduced primary production in the lake, however, as Secchi disc light penetration has increased since 1974–75 (Robarts, 1979; R. D. Robarts, this volume; Hydrobiology Research Unit, University of Zimbabwe, unpublished data).

The concentrations of the inorganic ions have, with few exceptions, followed the same trends as the nutrients and have decreased since the diversion of wastewaters to the irrigations schemes. Conductivity, being representative of the mineral status of the lake waters (Marshall and Falconer, 1973a), reached its maximum mean annual value of 17.2 mS m^{-1} during 1968–69. Since then, conductivity has decreased with values from 1976 onwards ranging between 10.1 and 12.6 mS m^{-1}. These values are somewhat higher than those of 8.8 and 9.5 mS m^{-1} recorded by Van der Lingen (1960) and Munro (1966) between 1957 and 1963. pH values have also decreased substantially since the late 1960s, although values over 9.0 are still recorded during algal blooms. Similarly, the concentrations of the major cations (sodium, calcium, magnesium, and potassium) have returned to a level similar to that reported by Marshall and Falconer (1973b) for the period 1960 to 1964, and iron concentrations have decreased by nearly an order of magnitude between 1969 and 1979 (Fig. 10). On the other hand, manganese concentrations have appeared to be relatively constant (Fig. 11). The anions (chloride and sulphate) seem to be another exception to the general trend shown by the other inorganic ions. The concentrations of the anions appear to have remained at relatively constant levels since 1968, although lack of data precludes a valid assessment. The reasons for this are unclear, but may relate to the inability of pasture lands to remove these ions.

It is interesting to note that the improvements observed in the water quality indices of Lake McIlwaine have had little effect on the formation of chemical gradients within the water column of the lake. The chemoclines noted by previous workers (Munro, 1966; Marshall and Falconer, 1973a; Mitchell and Marshall, 1974) still persist, albeit in most cases at much reduced intensities (Thornton, 1979a, 1980a, 1980b). This is particularly true for the oxycline which still leads to extensive de-oxygenation of the hypolimnion during the hot spring and summer months (Figs. 1 and 3). This hypolimnetic de-oxygenation during summer is a feature of most southern African man-made lakes (Allanson and Gieskes, 1961; Mitchell and Marshall, 1974; Walmsley and Toerien, 1977, 1979; Walsmley et al., 1978a, 1978b; Scott et al., 1977) and is not necessarily an indication of eutrophication as it is in temperate lakes (Mitchell and Marshall, 1974; Walmsley and Toerien, 1977; Wetzel, 1975).

There is little doubt that the diversion of nutrient-rich effluents from the Lake McIlwaine inflows has been effective in improving the water quality of the lake. This can be seen not only in the physico-chemical indices given in Table 2 but also in the reduction of the amount of phosphorus entering the lake (Table 1) from 288.1 metric tonnes of SRP-phosphorus in 1967 (Marshall and Falconer, 1973b) to a mean of 86.7 metric tonnes of TRP-phosphorus during 1976–78 (Thornton, 1979a, 1979b, 1980a). This is a reduction in mean phosphorus loading of over 70%, similar to the observed reduction of the phosphorus concentration in the lake. Ammonia nitrogen loading decreased by over 90%, whilst nitrate nitrogen, on the other hand, entered the lake in marginally higher amounts (Table 1).

This apparent lack of effect of pasture irrigation, tertiary treatment methods on nitrate has been shown to be a function of the type of crop produced in the pastures (J. McKendrick, this volume). In addition, nitrogen is difficult to control or remove with present-day wastewater treatment methods, and atmospheric – water nitrogen exchange is not unknown (Toerien, 1977; Hemens et al., 1977). These factors, combined with the fact that most southern African man-made lakes are naturally phosphorus limited (Toerien et al., 1975; Robarts and Southall, 1977), tend to suggest that phosphorus is of over-riding importance in determining lake trophic status, and hence the emphasis in the remainder of this section will be on phosphorus.

It may be suggested that the reduction in phosphorus concentration in Lake McIlwaine during 1977 is due to the more complete flushing of the lake during that year than during the pre-diversion study in 1967. Flow has been shown to play an important role in the phosphorus cycle (Dillon, 1975), with phosphorus concentration and flow in Lake McIlwaine being inversely related at low flows and directly related at high flows (Thornton, 1979a, 1980a). Thus, it is of some importance to compare years of equal flow when assessing the results of phosphorus diversion. To do this, Thornton (1980a) employed the method of Welch (1977) whereby a mean annual riverine phosphorus concentration and the long-term mean total annual river flow, calculated from the mean annual run-off (B. R. Ballinger and J. A. Thornton, this volume), are used to derive a phosphorus loading based on equivalent flows. Phosphorus loadings calculated by this method are shown in Table 3.

Table 3 shows an even greater reduction in the phosphorus loading than was previously indicated on the basis of a straight comparison of the data in

Table 3 Flow-normalised phosphorus loads in metric tonnes P derived from mean phosphorus concentrations (mg P l^{-1}) and mean river flows (10^6 m^3) after the method of Welch (1977)

Year	Mean P conc inflow	Mean annual inflow	Phosphorus loads per annum			% reduction from 1967
			m tonnes	g m^{-2}	g m^{-3}	
1967	2.25	304.7	685.6	26.1	2.8	—
1977	0.19	304.7	57.9	2.2	0.2	91
1978	0.13	304.7	39.6	1.5	0.2	94

Table 1. A reduction in loading of over 90% of the pre-diversion flow-normalised loads is seen. In both cases, it should be borne in mind that the data for 1967 are for the SRP fraction only, and that the equivalent reduction of the TRP-phosphorus load would be even greater.

Another approach to establishing the effectiveness of the nutrient diversion programme is to calculate the potential phosphorus load diverted from the lake to the pasture irrigation schemes. This potential load was calculated by Thornton (1980a) using total annual wastewater volumes and an estimated mean phosphorus concentration for secondarily treated effluent (10 mg P l^{-1}; Salisbury Sewerage Disposal Environmental Impact Statement Committee, 1979). These data are shown in Table 4 for the pre-diversion and post-diversion periods. Following diversion, no further effluent was intentionally released into the lake, and phosphorus in the run-off from the pasture irrigation schemes was minimal (W. K. Nduku, unpublished data; J. McKendrick, this volume). Although loads calculated by this method do not balance with the measured phosphorus loads in the rivers (Table 5), thus failing to provide reliable data, this indirect approach does allow an independent estimate of the

Table 4 Phosphorus loads to Lake McIlwaine in metric tonnes P calculated from total annual wastewater volumes (10^6 m^3)

Year	Parameter	Enters works	Discharge to irrigation	Discharge to rivers	P-load to lake (Table 1)	% P sewage
1967	P	156	nil	156	288	54
	flow	20	nil	20		
1977	P	289	262	27*	82	33
	flow	31	26	5		
1978	P	312	283	29*	92	32
	flow	33	29	4		

* Wet weather spillage.

Table 5 Estimated phosphorus loads to the Makabusi and Marimba Rivers in metric tonnes P derived from sewage and other sources compared to the actual measured loads

Year	Discharge to rivers	Discharge from other sources	Totals	Total P-load (from Table 1)
1967	156	4	160	285
1977	27	9	36	50
1978	29	18	47	60

magnitude of the diverted mass of phosphorus to be made. Based on a comparison of the data from 1967 with a mean of the data from 1976–78, there has been a reduction in the annual mass of phosphorus released into the rivers of 128 metric tonnes or over 70%; this is despite increases in the volume of sewage being treated. This value shows good agreement with that calculated using the field data.

The year to year variability in the study of Thornton (1979a, 1979b, 1980a) was low despite differences of flow of about 50% between 1976–77 and 1977–78. The majority of the difference in loading can be accounted for by the increased load observed in the Makabusi River, which suggests that phosphorus sources in that river's catchment are non-conservative. In the Hunyani and Marimba Rivers, the constancy of the phosphorus loading figures suggests a conservative source. The non-conservative nature of the phosphorus supply in the Makabusi River catchment is probably due to man-induced perturbations of the natural phosphorus supply, or to urban stormwater run-off (J. A. Thornton and W. K. Nduku, this volume). It is both interesting and disturbing to note the increasingly important role played by the latter component and related diffuse sources of phosphorus in the nutrient budget of Lake McIlwaine (Tables 4 and 5; Thornton, 1980c). Whilst major decreases in the total load of phosphorus have been achieved through the control of point sources and the diversion of municipal wastewater to irrigation schemes, these decreases have only affected the Makabusi and Marimba Rivers. A ten-fold increase in the phosphorus load in the Hunyani River has been observed since 1967 (Table 1) which is due predominantly to increased diffuse source run-off from continued urban development in the river catchment (J. A. Thornton and W. K. Nduku, this volume). Such increases, if continued unabated, will tend to negate the beneficial effects of the nutrient diversion scheme. This highlights the need for a catchment-based management approach to lake restoration.

References

Allanson, B. R. and J. M. T. M. Gieskes, 1961. Investigations into the ecology of polluted waters in the Transvaal. Part II. An introduction to the limnology of Hartbeespoort Dam with special reference to the effect of industrial and domestic pollution. Hydrobiol., 18: 77–94.

Dillon, P. J., 1975. The phosphorus budget of Cameron Lake, Ontario: The importance of flushing rate to the degree of eutrophy of lakes. Limnol. Oceanogr., 20: 28–39.

Falconer, A. C., B. E. Marshall and D. S. Mitchell, 1970. Hydrobiological studies of Lake McIlwaine in relation to its pollution, 1968 and 1969. University of Rhodesia Rep., Salisbury.

Grobbelaar, J. U., P. C. Keulder and P. Stegmann, 1980. Some properties of suspended sediment adsorbed cations in turbid freshwaters of South Africa. J. Limnol. Soc. Sth. Afr., 6: 55–58.

Hemens, J., D. E. Simpson and R. J. Warwick, 1977. Nitrogen and phosphorus input to the Midmar Dam, Natal. Water SA, 3: 193–201.
Marshall, B. E. and A. C. Falconer, 1973a. Physico-chemical aspects of Lake McIlwaine (Rhodesia), a eutrophic tropical impoundment. Hydrobiol., 42: 45–62.
Marshall, B. E. and A. C. Falconer, 1973b. Eutrophication of a tropical African impoundment (Lake McIlwaine, Rhodesia). Hydrobiol., 43: 109–124.
Mitchell, D. S. and B. E. Marshall, 1974. Hydrobiological observations on three Rhodesian reservoirs. Freshwat. Biol., 4: 61–72.
Munro, J. L., 1966. A limnological survey of Lake McIlwaine, Rhodesia. Hydrobiol., 28: 281–308.
Nduku, W. K., 1976. The distribution of phosphorus, nitrogen and organic carbon in the sediments of Lake McIlwaine, Rhodesia. Trans. Rhod. Scient. Ass., 57: 45–60.
Nduku, W. K., 1978. The thermocline stability in Lake McIlwaine, Rhodesia. Paper presented at the Limnological Society of Southern Africa Congress, Pietermaritzburg, Republic of South Africa.
Robarts, R. D., 1979. Underwater light penetration, chlorophyll a and primary productivity in a tropical African man-made lake (Lake McIlwaine, Rhodesia). Arch. Hydrobiol., 86: 423–444.
Robarts, R. D. and D. S. Mitchell, 1976. Management of highly productive dams. In: G. G. Cillie, ed., Proceedings of workshop on mineral enrichment and eutrophication of water. First Interdisciplinary Conference on Marine and Freshwater Research in Southern Africa, Port Elizabeth, Republic of South Africa. CSIR Spec. Rep. No. S122, Pretoria.
Robarts, R. D. and G. C. Southall, 1975. Algal bioassays of two tropical Rhodesian reservoirs. Acta Hydrochim. Hydrobiol., 3: 369–377.
Robarts, R. D. and G. C. Southall, 1977. Nutrient limitation of phytoplankton growth in seven tropical man-made lakes with special reference to Lake McIlwaine, Rhodesia. Arch. Hydrobiol., 79: 1–35.
Robarts, R. D. and P. R. B. Ward, 1978. Vertical diffusion and nutrient transport in a tropical lake (Lake McIlwaine, Rhodesia). Hydrobiol., 59: 213–221.
Salisbury Sewerage Disposal Environmental Impact Statement Committee, 1979. Report on Salisbury's sewerage disposal. Ministry of Water Development Rep., Salisbury.
Scott, W. E., M. T. Seaman, A. D. Connell, S. I. Kohlmeyer and D. F. Toerien, 1977. The limnology of some South African impoundments. I. The physico-chemical limnology of Hartbeespoort Dam. J. Limnol. Soc. Sth. Afr., 3: 43–58.
Talling, J. F. and I. B. Talling, 1965. The chemical composition of African lake water. Int. Revue ges. Hydrobiol., 50: 421–463.
Thornton, J. A., 1979a. Some aspects of the distribution of reactive phosphorus in Lake McIlwaine, Rhodesia: phosphorus loading and seasonal responses. J. Limnol. Soc. Sth. Afr., 5: 33–38.
Thornton, J. A., 1979b. Some aspects of the distribution of reactive phosphorus in Lake McIlwaine, Rhodesia: phosphorus loading and abiotic responses. J. Limnol. Soc. Sth. Afr., 5: 65–72.
Thornton, J. A., 1980a. Factors influencing the distribution of reactive phosphorus in Lake McIlwaine, Zimbabwe. D.Phil. Diss., University of Zimbabwe.
Thornton, J. A., 1980b. A comparison of the summer phosphorus loadings to three Zimbabwean water-supply reservoirs of varying trophic states. Water SA, 6: 163–170.
Thornton, J. A., 1980c. The Water Act, 1976, and its implications for water pollution control: case studies. Trans. Zimbabwe Scient. Ass., 60: 36–45.

Toerien, D. F., 1977. A review of eutrophication and guidelines for its control in South Africa. CSIR/NIWR Spec. Rep. No. WAT48, Pretoria.

Toerien, D. F., K. L. Hyman and M. J. Bruwer, 1975. A preliminary trophic state classification of some South African Impoundments. Water SA, 1: 15–23.

Van der Lingen, M. I., 1970. Some observations on the limnology of water storage reservoirs and natural lakes in Central Africa. First Fed. Sci. Congress Proc., 1–5.

Walmsley, R. D. and D. F. Toerien, 1977. The summer conditions of three eastern Transvaal reservoirs and some considerations regarding the assessment of trophic status. J. Limnol. Soc. Sth. Afr., 3: 27–41.

Walmsley, R. D. and D. F. Toerien, 1979. A preliminary limnological study of the Buffelspoort Dam and its catchment. J. Limnol. Soc. Sth. Afr., 5: 51–58.

Walmsley, R. D., D. F. Toerien and D. J. Steijn, 1978a. Eutrophication of four Transvaal dams. Water SA, 4: 61–75.

Walmsley, R. D., D. F. Toerien and D. J. Steijn, 1978b. An introduction to the limnology of Roodeplaat Dam. J. Limnol. Soc. Sth. Afr., 4: 35–52.

Watts, C. J., 1980. Seasonal variation of nutrient limitation of phytoplankton growth in the Hunyani River system with particular reference to Lake McIlwaine, Zimbabwe. M.Phil. thesis, University of Zimbabwe.

Welch, E., 1977. Nutrient diversion: resulting lake trophic state and phosphorus dynamics. USEPA Pub. No. EPA-600/3-77-003, Corvallis.

Wetzel, R. G., 1975. Limnology. Saunders, Philadelphia.

The sediments
R. Chikwanha, W. K. Nduku and J. A. Thornton

Sediment chemistry
J. A. Thornton and W. K. Nduku

Numerous studies of nutrient cycling in temperate lakes have shown the importance of the sediments as a source and/or sink of nitrogen and phosphorus (Golterman, 1977a). Previous investigations of Lake McIlwaine have also suggested the existence of a sediment source/sink of nutrients in that lake (Falconer et al., 1970; Marshall and Falconer, 1973), and calculations presented by Robarts and Ward (1978) have suggested that internal nutrient loading from this source might be considerable. More recently, Thornton (1979, 1980) has shown that sediment-water exchange processes do in fact have a significant effect on the lake nutrient budget (see J. A. Thornton and W. K. Nduku, this volume), but that the sediments act predominantly as a nutrient sink. This sink is shown in the high sediment nutrient concentrations measured by Nduku (1976).

The sediments of Lake McIlwaine are relatively high in nutrients (Table 6) and are indicative of a eutrophic lake (W. K. Nduku, unpublished data; Nduku and Robarts, 1977); non-eutrophic impoundments in Zimbabwe have

Table 6 Sediment chemical composition of Lake McIlwaine and other lakes (after Blair & Bowser, 1978). Concentrations in ppm except where noted.

Lake	Location	No. samples	P	N	Organic C	Ca	Mg	K	Na	Reference
McIlwaine	Zimbabwe	7	0.08–1.69	—	—	7–16	3–7	0.9–2.05	92–525	Falconer et al., 1970
		2	3.87	10.61	16%	0.96	0.23	0.31	0.15	Nduku & Robarts, 1977
		14	0.01–0.43	0.01–0.11	0.5–13%	0.2–1.6	0.08–0.6	0.05–0.3	0.02–0.17	Nduku, 1976
Umgusa	Zimbabwe	2	1.52	3.67	9%	—	—	0.57	1.14	Nduku & Robarts, 1977
Connemara No. 3	Zimbabwe	2	1.64	7.90	26%	0.19	0.04	0.06	0.05	Nduku & Robarts, 1977
Connemara No. 2	Zimbabwe	2	1.80	11.65	30%	0.19	0.04	0.12	0.07	Nduku & Robarts, 1977
George	Uganda	5	0.8–3.0	0.5–3.5	0–35%	0.5–2.0	0.1–1.7	—	—	Viner, 1977
Kivu	Uganda	381	—	—	—	9.5	1.2	—	—	Blair & Bowser, 1978
Tanganyika	Tanzania	31	—	—	—	1.2	1.2	—	—	Blair & Bowser, 1978
Edward	Uganda	51	—	—	—	3.0	1.2	—	—	Blair & Bowser, 1978
Albert	Uganda	92	—	—	—	1.4	1.4	—	—	Blair & Bowser, 1978
Ontario	Canada	2	0.07–0.22	—	—	0.4–6.4	1.3–2.0	2.3–3.2	0.5–1.9	Blair & Bowser, 1978
Erie	Canada	2	0.06–0.19	—	—	0.4–3.6	1.3–2.6	2.2–3.2	0.4–0.9	Blair & Bowser, 1978
Michigan	Canada	2	0.02–0.13	—	2.01	3.2–10.8	1.4–4.1	0.2–1.2	0.03–0.3	Blair & Bowser, 1978
Superior	Canada	3	0.14	—	2.3–5.4	1.2–2.4	2.3	0.5–2.4	1.25	Blair & Bowser, 1978
Linsley	USA	1	—	—	—	0.9	0.6	—	—	Blair & Bowser, 1978
Mendota	USA	2	0.1–0.2	—	9.5–9.9	22.8	1.65	0.5–1.2	1.8	Blair & Bowser, 1978
Wingra	USA	1	0.05	—	6.9	23.8	0.6	0.07	—	Blair & Bowser, 1978
Little St. Germain	USA	1	1.4	—	23.5	0.12	0	0.36	—	Blair & Bowser, 1978
Trout	USA	1	0.6	—	22.3	0.4	0.4	0.5	—	Blair & Bowser, 1978
Minocqua	USA	1	0.7	—	18.1	0.3	0.4	0.9	—	Blair & Bowser, 1978
Weber	USA	1	0.3	—	31.7	0.1	0.2	0.6	—	Blair & Bowser, 1978
Little John	USA	1	0.2	—	33.3	0.3	0.4	0.5	—	Blair & Bowser, 1978
Devils	USA	2	0.1	—	12.9	0.2–0.5	0.4–1.4	tr–0.9	tr	Blair & Bowser, 1978
Monoma	USA	3	0.1–0.2	—	9.7	10.1–19.7	1.1–1.3	0.07–0.88	—	Blair & Bowser, 1978
Crystal	USA	1	—	—	—	1.7	0.9	—	—	Blair & Bowser, 1978

sediment phosphorus concentrations of about 0.3 ppm of phosphorus whilst eutrophic lakes have concentrations in excess of 1.0 ppm. Lake McIlwaine sediments have phosphorus concentrations which span this range, ranging from 0.01 to 3.87 ppm (Table 6), with the higher concentrations being recorded in deeper water (Nduku, 1976). This suggests that sediment-bound phosphorus is related to sediment particle size. Smaller-sized particles will not settle out of solution as rapidly as larger particles in response to the decreased flow velocity as the riverine inflow enters the lake and thus will accumulate in the deeper portions of the impoundment. This has been confirmed by Nduku (1976) who showed a direct relationship between clay content of the sediments and the amount of bound phosphorus, both increasing in deeper water. Nitrogen followed similar trends (Nduku, 1976).

In addition, sediment nutrient concentrations were found to be related to the percentage of organic carbon in the sediments (Nduku, 1976) as organic matter also tends to accumulate in the deeper portions of the lake basin. Nduku (1976) records that at the time of his study (1974) much of the organic matter in the lake was derived from allochthonous sources, in particular from the Salisbury wastewater which had been discharged into the influent rivers to the lake and from terrestrial litter carried into the lake during the summer floods. Anoxic conditions in the hypolimnion of the lake resulted in little degradation of the organic matter and lack of wave action allowed it to accumulate. Examination of phosphorus and nitrogen concentrations in the epilimnetic sediments showed relatively low concentrations; these areas of the lake ($z < 10$ m) tend to be well oxygenated and subject to biological perturbations by oligochaetes and chironomids (see B. E. Marshall, this volume), and wind-induced re-suspension of the sediments. Further, sediments from shallow areas of Lake McIlwaine can be sandy and hence there are fewer potential binding sites for nitrogen and phosphorus radicals than in the deeper water areas (Nduku, 1976; Thornton, 1979, 1980).

The foregoing also suggests that sediment nutrient concentrations are related to oxygen concentrations in the overlying water and to depth. Nduku (1976) showed that these correlations exist, and in particular stressed the effects of anoxic conditions in the hypolimnion of Lake McIlwaine on the release of reactive phosphorus and ammonia nitrogen, and on the accumulation of some cations (specifically, calcium, magnesium, sodium and potassium). He has also shown that these cations, which play a part in the diagenetic formation of clay minerals, can act to bind reactive phosphorus and ammonia fractions more strongly to the sediments and reduce the magnitude of the sediment-water exchange process. Similar correlations have been observed elsewhere (Clay and Wilhm, 1979).

Interstitial water chemistry
Few measurements of the interstitial or pore water nutrient concentrations of Lake McIlwaine have been made. Nduku (unpublished) used various techniques during 1976 for extracting pore water from sediment samples (squeezing and centrifugation) and measured the nitrate and ammonia nitrogen concentrations and reactive phosphorus concentrations of these samples. Nitrate and ammonia ranged from 0.07 to 0.26 mg NO_3-N l^{-1} and from 19.2 to 50 mg NH_3-N l^{-1} respectively. Reactive phosphorus ranged from 0.027 to 0.828 mg P l^{-1} with a mean of 0.535 mg l^{-1}. Thornton (1979, 1980) using a dialysis technique (Mayer, 1976) found a slightly higher concentration of reactive phosphorus in the lake sediments during 1978; mean values at three lake stations (stations SM-4, SM-3 and SM-1 in Fig. 2 of J. A. Thornton and W. K. Nduku, this volume) were 1.16, 1.05 and 1.13 mg P l^{-1} respectively. Station SM-4 is the same as that used by Nduku. The difference between the phosphorus concentrations observed by Nduku and Thornton (1979, 1980) may be related to the methodology involved; however, it may have real environmental significance and be related to the decreased phosphorus concentrations in the overlying waters which would allow greater resolubilisation of bound phosphorus forms and lead to the increased internal phosphorus loading predicted by Robarts and Ward (1978). Viner (1975a) reported a similar range of concentrations in the interstitial water of Lake George, Uganda.

Sediment-water exchange
Robarts and Ward (1978) have calculated the vertical mass-transport of reactive phosphorus and ammonia in Lake McIlwaine as a function of vertical diffusivity. Using a mean vertical diffusivity coefficient, K_z, of 0.21 cm^2 s^{-1} they estimated a vertical mass-transport of the nutrients amounting to 55 mg P m^{-2} d^{-1} and 145 mg NH_3-N m^{-2} d^{-1} at the assumed sediment-water interface. Subsequently, Thornton (1979, 1980) measured the sediment-water exchange process using a modified bell-jar technique (Lee, 1977). Rates of net sediment-water exchange measured during this latter study averaged + 4.2 mg P m^{-2} d^{-1}, − 14.3 mg NO_3-N m^{-2} d^{-1} and + 4.7 mg NH_3-N m^{-2} d^{-1} (the signs indicate a net flux from the sediments, +, and to the sediments, −). These rates suggest that other processes are at work within the sediments of the lake not accounted for by the diffusion processes as suggested by Robarts and Ward (1978). The net nutrient flux rates of Lake McIlwaine sediment are comparable to those measured elsewhere in the world (Silberbauer, 1981; Viner, 1975a; Fillos and Swanson, 1975; Kamp-Nielsen, 1974; Moller-Anderson, 1974).

Thornton (1979, 1980) has investigated the aerobic release of phosphorus from Lake McIlwaine sediments in the laboratory. The release rates observed during these experiments were between 0.03 and 0.13 mg P l^{-1} d^{-1} (the initial concentration of phosphorus in the overlying water ranged from 0.01 to 0.10 mg P l^{-1}). The observed in-lake phosphorus release rate of 4.2 mg P m^{-2} d^{-1} (or 0.05 mg P l^{-1} d^{-1}) is within the range observed in the laboratory. The laboratory release rates also compare favourably with those measured in other African lakes. Aerobic phosphorus release in Midmar Dam (South Africa) ranged from 0.02 to 0.10 mg P l^{-1} for an initial concentration of between 0.025 and 0.10 mg P l^{-1} (Furness and Breen, 1978). In Lake George (Uganda) phosphorus release was slightly higher, ranging from 0.28 to 0.44 mg P l^{-1} in samples having an initial concentration of 0.04 mg P l^{-1} (Viner, 1975a), but this may be a function of the nature of the sediments in that lake. Lake George sediments are highly organic (Viner, 1975b) compared with the drowned terrestrial soils of Lake McIlwaine (Thornton, 1979, 1980) and Midmar Dam (Furness and Breen, 1978).

The laboratory experiments of Thornton (1979, 1980) also showed that the sediments of Lake McIlwaine removed phosphorus from the water column, although Thornton (1979, 1980) suggests that in nature this uptake of phosphorus is most probably a function of adsorption onto suspended sediments in the influent rivers with subsequent deposition of the bound phosphorus in the lake. Uptake rates recorded under aerobic conditions in the laboratory ranged from 0.54 mg P g^{-1} to 0.93 mg P g^{-1} dry mass of sediment, or between 54 and 93% of the phosphorus supplied. The estimated percentage uptake of phosphorus observed in the lake (derived from the phosphorus mass-balance) is 55% and lies within this range (see Table 6, J. A. Thornton and W. K. Nduku, this volume). Similarly, the magnitude of the phosphorus removal process in Lake McIlwaine may be compared favourably with that of Midmar Dam, where Twinch and Breen (1978) recorded the uptake of 90% of the phosphorus supplied. Phosphorus uptake in Lake George, on the other hand, amounted to only 2% of the phosphorus supplied (Viner, 1975b).

In terms of the net effect of the sediments on the nutrient budget of Lake McIlwaine, Thornton (1979, 1980) has calculated a nutrient budget based on the field measurements of phosphorus uptake and release. Table 7 summarises his results and shows clearly that there is a net loss of phosphorus from the water column of the lake. These data also suggest that much of the phosphorus lost from the water column in this way remains locked up in the sediments with only limited exchange taking place. This net removal of phosphorus from the water column has been shown to occur in other lakes in both the temperate and tropical zones (Viner, 1975a, 1975b; Nduku and Robarts, 1977; Schindler *et al.*, 1977; Ahlgren, 1977).

Table 7 Phosphorus loads to Lake McIlwaine during 1977–78 showing the magnitude of sediment-water exchange processes in relation to the annual phosphorus load to the lake

Source	Load (g P m^{-2} a^{-1})
P-load t the lake from external sources*	3.9
P-load to the lake from the sediments	0.6
Total P-load	4.5
P-loss to spillway and algae	1.2
P-loss to sediments	3.3
Total P-loss	4.5

* Table 6, J. A. Thornton and W. K. Nduku, this volume.

Biological availability of sediment phosphorus

Thornton (1979, 1980) conducted batch-culture bioassays using the method of Watts (1980; this volume) and Robarts and Southall (1977) but substituting 1 g of air-dried sediment for the phosphorus requirement of the media (Golterman, 1977b). An indigenous culture of *Microcystis aeruginosa* Kutz was used as the assay organism. Good growth was obtained in the bioassays with better growth being obtained under slightly basic conditions (pH = 8.5) than under acidic conditions (pH = 6.5). Algal growth potentials (AGP's; Toerien *et al.*, 1975) measured in the bioassays averaged 290 mg l^{-1} and 55 mg l^{-1} dry biomass of *Microcystis* under basic and acidic conditions respectively. AGP's measured in non-limited lake bioassays averaged between 120 and 240 mg l^{-1} depending on the season (Thornton, 1980; Watts, 1980, this volume). Thus, it may appear that the sediment-water phosphorus exchange process could potentially supply enough of the nutrient to sustain the aquatic flora of Lake McIlwaine. This is in contrast to observations made by Viner (1975b) on Lake George, but is consistent with the suggestions made by Golterman (1977b) and Grobler and Davies (1979; 1981). Nevertheless, Table 7 shows that in the lake the conditions governing uptake and release of phosphorus from the lake sediments are such that sediment phosphorus release is minimal and thus its influence on the algal populations may also be considered to be minimal.

Conclusions

In conclusion therefore it would appear that despite the high concentrations of nitrogen and phosphorus in the lake sediments the sediments act predominantly as a nutrient sink and should only enhance the recovery of the lake now that the nutrient diversion programme has been fully implemented.

References

Ahlgren, I., 1977. Role of sediments in the process of recovery of a eutrophied lake. In: H. L. Golterman, Interactions between sediments and freshwater. Junk, The Hague.
Blair, F. and C. J. Bowser, 1978. The mineralogy and related chemistry of lake sediments. In: A. Lerman, Lakes: Chemistry, geology and physics. Springer Verlag, New York.
Clay, E. M. and J. Wilhm, 1979. Particle size, percent organic carbon, phosphorus, mineralogy and deposition of sediments in Ham's and Arbuckle Lakes. Hydrobiol., 65: 33–38.
Falconer, A. C., B. E. Marshall and D. S. Mitchell, 1970. Hydrobiological studies of Lake McIlwaine 1968–69. University of Rhodesia Rep.
Fillos, J. and W. R. Swanson, 1975. The release rate of nutrients from river and lake sediments. J. Wat. Pollut. Control Fed., 47: 1032–1042.
Furness, H. D. and C. M. Breen, 1978. The influence of P-retention by soils and sediments on the water quality of the Lions River. J. Limnol. Soc. Sth. Afr., 4: 113–118.
Golterman, H. L., 1977a. Interactions between sediments and freshwater. Junk, The Hague.
Golterman, H. L., 1977b. Sediments as a source of phosphate for algal growth. In: H. L. Golterman, Interactions between sediments and freshwater. Junk, The Hague.
Grobler, D. C. and E. Davies, 1979. The availability of sediment phosphate to algae. Water SA, 5: 114–122.
Grobler, D. C. and E. Davies, 1981. Sediment as a source of phosphates: a study of 38 impoundments. Water SA, 7: 54–60.
Kamp-Nielson, L., 1974. Mud-water exchange of phosphate and other ions in undisturbed sediment cores and factors affecting the exchange rates. Arch. Hydrobiol., 73: 218–237.
Lee, D. R., 1977. A device for measuring seepage flux in lakes and estuaries. Limnol. Oceanogr., 22: 140–147.
Lerman, A., 1978. Lakes: Chemistry, geology and physics. Springer Verlag, New York.
Marshall, B. E. and A. C. Falconer, 1973. Physico-chemical aspects of Lake McIlwaine (Rhodesia), a eutrophic tropical impoundment. Hydrobiol., 42: 45–62.
Mayer, L. M., 1976. Chemical water sampling in lakes and sediments with dialysis bags. Limnol. Oceanogr., 21: 909–912.
Moller-Anderson, J., 1974. Nitrogen and phosphorus budgets and the role of sediments in six shallow Danish lakes. Arch. Hydrobiol., 74: 528–550.
Nduku, W. K., 1976. The distribution of phosphorus, nitrogen and organic carbon in the sediments of Lake McIlwaine, Rhodesia. Trans. Rhod. Scient. Ass., 57: 45–60.
Nduku, W. K. and R. D. Robarts, 1977. The effect of catchment geochemistry and geomorphology on the productivity of a tropical African montane lake. Freshwat. Biol., 7: 19–30.
Robarts, R. D. and G. C. Southall, 1977. Nutrient limitation of phytoplankton growth in seven tropical man-made lakes with special reference to Lake McIlwaine, Rhodesia. Arch. Hydrobiol., 79: 1–35.
Robarts, R. D. and P. R. B. Ward, 1978. Vertical diffusion and nutrient transport in a tropical lake (Lake McIlwaine, Rhodesia). Hydrobiol., 59: 213–221.
Schindler, D. W., R. Hesslein and G. Kipphut, 1977. Interactions between sediments and overlying waters in an experimentally eutrophied Pre-Cambrian Shield lake. In: H. L. Golterman, Interactions between sediments and freshwaters. Junk, The Hague.

Silberbauer, M. J., 1981. Laboratory and lake measurements of phosphate exchange between the sediments and monimolimnion of Swartvlei. Paper presented at the Limnological Society of Southern Africa Congress, Bloemfontein, Republic of South Africa.

Thornton, J. A., 1979. Some aspects of the distribution of reactive phosphorus in Lake McIlwaine, Rhodesia; phosphorus loading and abiotic responses. J. Limnol. Soc. Sth. Afr., 5: 65–72.

Thornton, J. A., 1980. Factors influencing the distribution of reactive phosphorus in Lake McIlwaine, Zimbabwe. D.Phil. Diss., University of Zimbabwe.

Toerien, D. F., K. L. Hyman and M. J. Bruwer, 1975. A preliminary trophic status classification of some South African impoundments. Water SA, 1: 15–23.

Twinch, A. J. and C. M. Breen, 1978. Enrichment studies using isolation columns. II. The effects of phosphorus enrichment. Aquat. Bot., 4: 161–168.

Viner, A. B., 1975a. The sediments of Lake George (Uganda). II. Release of ammonia and phosphate from an undisturbed mud surface. Arch. Hydrobiol., 76: 368–378.

Viner, A. B., 1975b. The sediments of Lake George (Uganda). III. The uptake of phosphate. Arch. Hydrobiol., 76: 393–410.

Watts, C. J., 1980. Seasonal variation of nutrient limitation of phytoplankton growth in the Hunyani River system, with particular reference to Lake McIlwaine, Zimbabwe. M.Phil. thesis, University of Zimbabwe.

Sediment transport
R. Chikwanha

Lake McIlwaine has a catchment area of 2227 km^2. Figure 13 shows the whole catchment of the lake and the locations of the flow gauging stations referred to in this paper. It also shows the location of the only regular suspended sediment sampling station in the catchment (station C.21). The geology of the catchment is predominantly granite (see K. Munzwa, this volume) and the land use can be divided into two broad divisions; namely, Tribal Trust Land and Commercial Farming Land. The former is found mainly in the Nyatsime catchment, the main tributary catchment of the Hunyani River upstream of station C.21. The latter form of land use is confined mainly to the Hunyani catchment. However, it is important to note that these are broad generalisations (see K. Munzwa, this volume) and that neither of the two land use patterns is exclusive to the above areas. The Makabusi and Marimba Rivers, which drain the Salisbury City Centre and the southern section of Greater Salisbury, have an urban land use. These two rivers are particularly important as effluent channels.

Suspended sediment sampling on the Hunyani River started in December 1976 at station C.21 (Ward, 1977; Chikwanha, 1980a). Sampling has since been carried out annually during the rainy season using a Kahlsico POSAWS (Portable Sequential Aliquot Water Sampler). The use of the automatic

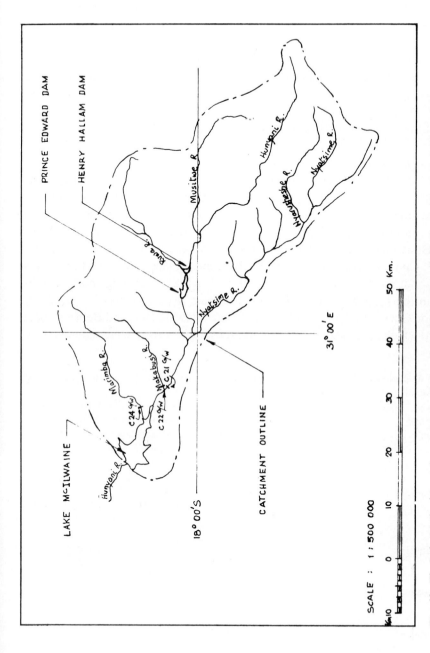

Fig. 13 The Lake McIlwaine catchment showing the locations of the sediment sampling stations C.21, C.22 and C.24.

sampler at station C.21 probably makes the sampling at this station more reliable than at any of the other stations in the catchment where samples are taken manually by resident observers. Sampling of the Makabusi and Marimba Rivers was primarily for hydrobiological purposes although it is from the occasional analysis of these samples for suspended sediment that estimates of the sediment loads carried by these rivers were obtained (Chikwanha and Ward, 1979). Sampling of the Nyatsime River was carried out during the 1977–78 season with the objective of establishing the proportion of the sediment load to Lake McIlwaine carried by the Nyatsime River.

Sediment yields

The sediment yields of the Hunyani River catchment, if calculated over the whole catchment, are deceptively low. Table 8 shows the sediment loads and the sediment yields for the Hunyani River and its tributaries for the period between December 1976 and December 1979. A comparison of the yields in the Hunyani catchment with the Umsweswe catchment (station C.87; Ward, 1977; Chikwanha and Ward, 1979; Chikwanha, 1980b) shows that the yields are about five times less than those of the Umsweswe although both stations are in the same hydrological zone. The main reason for this is that above station C.21 there are two reservoirs (Prince Edward Dam and Henry Hallam Dam) which act as sediment traps and substantially reduce the sediment load

Table 8 Sediment loads and yields for the Hunyani River and tributaries upstream of Lake McIlwaine

River and station	Season	Catchment area (km^2)	Total suspended load (excluding bed load) (10^3 tonnes)	Sediment yield (tonnes km^{-2})
Hunyani C.21	1976–77	1510	33.90	22.45
	1977–78		31.80	21.06
	1978–79		2.49	1.65
Makabusi C.22	1977–78	231	3.8	16.45
	1978–79		0.3	1.30
Marimba C.24	1977–78	189	4.7	24.87
	1978–79		0.4	1.96
Nyatsime C.23	1977–78	500	8.0	16.00

from the upper Hunyani catchment. Most of the larger particles are trapped in the two upstream reservoirs.

The sediment yield of the Nyatsime catchment is also surprisingly low (only 16 tonnes km^{-2} as compared to 21 tonnes km^{-2} for the whole of the Hunyani catchment) particularly if one considers that the catchment comprises some heavily populated areas of the Seki and Chiota Tribal Trust Lands. Table 8 also gives the estimated values of the suspended loads transported by the Makabusi and Marimba Rivers. These amount to about 12% of the total for the whole of the Hunyani catchment. While the sediment loads for the Marimba and Makabusi Rivers are not negligible, it is primarily in terms of nutrients (particularly nitrogen and phosphorus) that the role of these two rivers is most important (see J. A. Thornton and W. K. Nduku, this volume).

Sediment rating curves

For every season a logarithmic plot of the daily sediment loads against the corresponding daily run-offs was obtained; straight lines of best fit, calculated by the least squares method, gave the rating curve equations shown in Table 9 (Ward, 1977; Chikwanha and Ward, 1979; Chikwanha, 1980b). The table shows also the relative run-off ($Q / \bar{Q}$), the ratio of the total annual run-off (Q) and the long-term mean run-off ($\bar{Q}$). This ratio is a measure of the 'wetness' of the season. Generally the constant (B) of the rating curve equation increases as the relative run-off decreases. In a very wet season like that of 1977–78 the logarithmic plot of sediment load against run-off shows a distinct separation of the early season before the first major flood of the main season. The early season equation has a very high value of the constant (B). Dry seasons tend to follow the same pattern as that of the early part of a very wet season.

Siltation

For the purposes of determining how much of the incoming sediment was deposited in Lake McIlwaine sediment samples were also taken on the spillway. The calculation of the probable depths of deposition is based on the assumption that the sediment is evenly distributed over the whole reservoir bottom. A range of specific gravities of between 1.2 and 1.9 is assumed to represent the probable range of densities of deposition; the thickness of the deposition ranging between the two extremities of the range in densities is shown as a range in Table 10. The same table also shows the seasonal sediment inflows and outflows for Lake McIlwaine. The siltation results shown in Table 10 suggest a total loss of capacity of 0.22% for Lake McIlwaine

Table 9 Rating curve equations for sediment loads carried by the Hunyani River at station C.21; Q_s = sediment discharge in tonnes; Q = water discharge in 10^3 m³; $\bar{Q}$ = long-term mean water discharge in 10^3 m³; and B = constant

Season	Relative run-off $Q/\bar{Q}$	Description of the season	Rating curve equations $Q_s = B Q^a$
1976–77	1.67	Wet	$Q_s = 0.79\, Q^{1.48}$
1977–78	2.45	Very wet	Early season: $Q_s = 6.08\, Q^{1.08}$ Main season: $Q_s = 1.31\, Q^{1.25}$
1978–79	0.38	Dry	$Q_s = 1.23\, Q^{1.33}$

Table 10 Seasonal accumulation of sediment in Lake McIlwaine

Season	Suspended loads		Seasonal accumulation depth of deposition (mm)
	Entering	Leaving	
1976–77	49×10^3	6×10^3	1–5
1977–78	40×10^3	negligible	1–5
1978–79	3.2×10^3	nil	0.08–0.38

in the three seasons studied (Ward, 1980). Over a period of 27 years since the lake was formed the estimated loss of capacity has been 2% of the total original capacity. This is a very low figure and the reduction of capacity of Lake McIlwaine as a result of sediment deposition is minimal. However, the contribution of the sediment deposits to the eutrophication and recovery of the lake should not be readily brushed aside (see J. A. Thornton and W. K. Nduku, this volume).

References

Chikwanha, R., 1980a. Sediment research in Zimbabwe. Hydrological Branch Rep., Ministry of Natural Resources and Water Development, Salisbury.

Chikwanha, R., 1980b. Sediment yields from Rhodesian rivers, 1978–1979 season. Hydrological Branch Rep., Ministry of Natural Resources and Water Development, Salisbury.

Chikwanha, R. and P. R. B. Ward, 1979. Sediment yields from Rhodesian rivers, 1977–1978 season. Hydrological Branch Rep., Ministry of Natural Resources and Water Development, Salisbury.

Ward, P. R. B., 1977. Sediment yields from Rhodesian rivers, 1976–1977 season. Hydrological Branch Rep., Ministry of Natural Resources and Water Development, Salisbury.

Ward, P. R. B., 1980. Sediment transport and reservoir siltation formula for Zimbabwe Rhodesia. Civ. Engr. S. Afr., 22: 9–15.

Ward, P. R. B. and R. Chikwanha, 1980. Laboratory measurement of sediment by turbidity. J. Hydraul. Div. Am. Soc. civ. Engrs., 106 (HY6): 1041–1053.

The effects of urban run-off
R. S. Hatherly, W. K. Nduku, J. A. Thornton and K. A. Viewing

The aqueous phase: nutrients in run-off from small catchments
J. A. Thornton and W. K. Nduku

The effects of land use within riverine drainage basins upon the water quality of the river have been known intuitively and qualitatively for many years but have only recently been quantified (Dillon and Kirchner, 1975; Likens *et al.*, 1970, 1977). This has stemmed largely from water quality management requirements in terms of water pollution control legislation, and particularly from the need to be able to predict potential water quality problems arising from the development of catchment areas (Dillon and Rigler, 1974). The effects of catchment land use on the magnitude of nitrogen and phosphorus losses to water courses are of particular significance in view of the importance of these nutrients in determining lake trophic status (McColl *et al.*, 1975; Dillon and Rigler, 1974; Vollenweider, 1971). In Zimbabwe, where stringent water pollution control legislation has virtually eliminated point source inputs of polluting loads of nitrogen and phosphorus (see D. B. Rowe, this volume), diffuse source run-off is becoming a major factor in the creation and maintenance of water quality problems (Thornton, 1980a). The study of Thornton and Nduku (1982) was therefore designed as one phase of an on-going study of the effects of land use on catchment nutrient export.

The four stations in the Chitungwiza urban area sampled by Thornton and Nduku (1982) show the effect of various stages of urbanisation in a catchment area (Fig. 14), with the stage of development in each catchment ranging from a well-established high density residential area (St. Mary's) through newly established areas (Seke and Zengeza) to virtually undeveloped areas (Nyamapfupfu). The close proximity of these stations to each other allows direct comparison of the data. Generally, there were noticeable increases in the concentrations of the nitrogen fractions between the undisturbed areas and developed lands. Mean concentrations of the combined nitrogen forms at the Nyamapfupfu station were 0.255 mg N l^{-1} whilst the mean concentrations

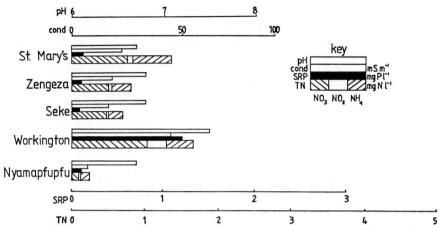

Fig. 14 Mean nutrient concentrations, pH and conductivity values recorded from the urban and industrialised catchments in the Lake McIlwaine area.

at the other three stations (St. Mary's, Seke and Zengeza) were 1.316 mg l⁻¹, 0.816 mg l⁻¹ and 0.687 mg l⁻¹ respectively (with an overall mean of 0.940 mg N l⁻¹). Nitrogen concentrations increased as the catchment areas became more developed with the St. Mary's station having the highest concentration and the partially developed Seke catchment having the lowest. Conductivity also increased from 8.0 mS m⁻¹ at the Nyamapfupfu station to upwards of 10.0 mS m⁻¹ at the other stations. Phosphorus concentrations and pH at the urban stations were little different from those recorded in the natural catchment. In each case, the urban run-off water quality lies within the guidelines set out in the Water (Effluent and Waste Water Standards) Regulations, 1977 (cf. Thornton, 1980a, Table 1; D. B. Rowe, this volume).

The industrial area of Workington did not have an undisturbed catchment in close enough proximity and hence direct comparison of the water quality of this catchment to a natural catchment was not possible. Nevertheless, it is clear from Fig. 14 that this catchment produced much higher concentrations of nitrogen and phosphorus, averaging 1.6 mg N l⁻¹ and 1.2 mg P l⁻¹, than any of the other catchment types. Increased concentrations of both nutrients as well as pH and conductivity are likely to have occurred in the Workington catchment when data from this catchment are compared with data from elsewhere in the Hunyani River catchment (Thornton, 1980b). Phosphorus concentrations at this station exceed the discharge standards given in the Water (Effluent and Waste Water Standards) Regulations, 1977.

The nutrient concentrations in the run-off from the urban and industrial catchments in the Greater Salisbury area are shown in Fig. 15 in comparison

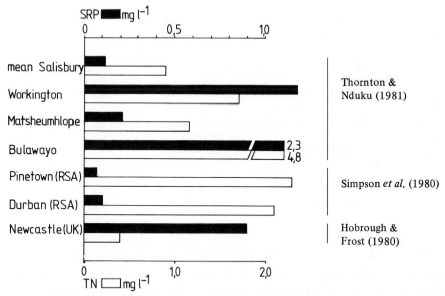

Fig. 15 Comparison of the mean nutrient concentrations from Zimbabwean catchments with similar catchments elsewhere.

with the concentrations observed in similar catchments elsewhere. The nutrient concentrations observed in the industrial catchments (Bulawayo and Workington) appear to be excessively high when compared to the urban (residential and commercial) watersheds both within Zimbabwe and outside of the country. This may be because of the fact that the urban catchments are not industrialised and hence are not strictly comparable. However, the comparison of these watersheds with the urbanised catchments does provide the necessary perspective from which to assess the relative magnitude of the nutrient contributions from industrialised catchments.

The other urban catchments shown in Fig. 15 (Durban, Pinetown and Newcastle) do have similar patterns of land use as the Salisbury catchments. The southern African watersheds appear more closely related to each other in terms of nutrient export than to the British and American catchments. This may be a function of geography but it is probably more likely to be due to the fact that run-off in the southern African catchments is nearly exclusively stormwater run-off (Simpson *et al.*, 1980; Simpson and Hemens, 1978; Thornton and Nduku, 1982) whilst the British and American catchments are influenced by domestic wastewater inputs (Hobrough and Frost, 1980;

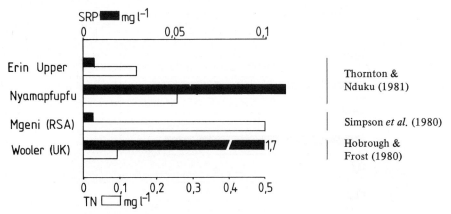

Fig. 16 Comparison of the mean nutrient concentrations from natural catchments in Zimbabwe with similar catchments elsewhere.

Omernik, 1976). The four southern African catchments, then, all show that nitrogen export exceeds phosphorus export in stormwater run-off in a ratio (N:P) of between 6:1 and 35:1 depending on the particular catchment. The higher ratio was observed in the Pinetown commercial district (Simpson *et al.*, 1980) whilst the lower was recorded in the low density residential area in the Matsheumhlope catchment (Thornton and Nduku, 1982).

The predominance of nitrogen in run-off from forested catchments reported by Thornton and Nduku (1982) is also reflected in the natural catchment studied (Fig. 16). An exception to this is the Wooler catchment (Hobrough and Frost, 1980). The higher phosphorus concentrations in this catchment as well as the somewhat higher concentrations of both nitrogen and phosphorus in the Nyamapfupfu catchment probably reflect the application of chemical fertilisers within the catchments.

Effects of diffuse source run-off

As flow data from the majority of the catchments studied by Thornton and Nduku (1982) were not available at the time of writing, they estimated flows in the Salisbury area based on an extrapolation of available data from the St. Mary's catchment and calculated the nutrient loads carried by the Salisbury streams on this basis. This allowed them to make an assessment of the effect of urban run-off on the nutrient budget of Lake McIlwaine. Fig. 17 shows the contributions of the three major influent rivers to the lake during 1967 and 1977 (Marshall and Falconer, 1973; Thornton, 1980b; J. A. Thornton and

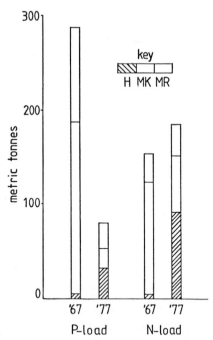

Fig. 17 Nutrient loads to Lake McIlwaine from the Hunyani (H), Makabusi (MK) and Marimba (MR) Rivers during 1967 and 1977 showing the increase in nutrient loads carried by the Hunyani following urbanisation of further portions of the catchment.

W. K. Nduku, this volume) and illustrates the reduction in nutrient loading that has taken place between those years. However, the greatest reduction in nutrient loading has been in terms of phosphorus loading whilst nitrogen loadings have remained virtually unaffected. Fig. 17 also shows that the contribution of the Hunyani River to both the nitrogen and phosphorus budgets of the lake has increased markedly. Thornton (1980a, 1980b) has suggested that this is due to the development within the Hunyani River catchment and particularly to the expansion of the Chitungwiza urban area which has led to increased loading from diffuse sources.

Table 11 shows the estimated nutrient loads carried by the study streams. From these estimates it is possible to calculate the theoretical nutrient budget of Lake McIlwaine as follows. Total flows at the four stations in the Hunyani River watershed (St. Mary's, Seke, Zengeza, and Nyamapfupfu) amounted to 3.4×10^6 m^3 or roughly 1% of the mean annual inflow volume to the lake (Thornton, 1980b). Assuming that the total nutrient load amongst the four Chitungwiza stations is representative of 1% of the catchment (as it is in terms of area), the total load to the lake would be on the order of 35 metric tonnes of

Table 11 Phosphorus and nitrogen loads (in kg) and catchment export rates (in mg m^{-2} a^{-1}) calculated using extrapolated flow values (in 10^3 m^3) for Salisbury area catchments (after Thornton and Nduku, 1982).

Parameter	St. Mary's	Zengeza	Seke	Nyamapfupfu	Workington
P-load	42	5	147	157	1255
N-load: NO_3-N*	260	26	774	146	1133
NO_2-N	21	2	34	21	281
NH_4-N	176	10	291	188	362
total	457	38	1099	355	1776
P-export	13	14	53	14	280
N-export	209	114	85	31	584
Flow	365	46	1600	1392	1077

* NO_3-N fraction used in calculations and Fig. 17.

phosphorus and 121 tonnes of nitrogen from the Hunyani River. Similarly, if the total flow estimated at the Workington station is assumed to be 5% of the total mean annual flow in the Marimba River, then the total load to the lake from that river would be on the order of 25 metric tonnes of phosphorus and 23 metric tonnes of nitrogen. These estimates agree extremely well with the measured loads during 1977–78 when river flows approximated the flows assumed above (Thornton, 1980b; J. A. Thornton and W. K. Nduku, this volume). This agreement between the predicted and observed values not only confirms the major role of diffuse source run-off in maintaining and creating eutrophication problems in Zimbabwe, but also suggests the usefulness of this sort of predictive capability as a management tool (Dillon and Rigler, 1975). However, further work is required on other catchment types before nutrient loads to water courses can be accurately predicted from land use data on the basis of catchment nutrient export values such as those given in Table 11.

The fact that diffuse source stormwater run-off can potentially supply sufficient nutrients to lakes such as Lake McIlwaine to maintain a eutrophic state is cause for concern, particularly when the continued expansion of urban centres such as Chitungwiza is considered. It suggests that despite the effective control of point source discharges through comprehensive water pollution control legislation Zimbabwean lakes may continue to be or become eutrophic (cf. Thornton, 1980a, 1980b). To prevent such occurrences in the future, it will be necessary to control, through legislation if necessary, the entry of stormwater run-off into natural water courses.

References

Dillon, P. J. and W. B. Kirchner, 1975. The effects of geology and land use on the export of phosphorus from watersheds. Wat. Res., 9: 135–148.

Dillon, P. J. and F. H. Rigler, 1974. A test of a simple nutrient budget model predicting the phosphorus concentration in lake water. J. Fish. Res. Bd. Can., 31: 1771–1778.

Dillon, P. J. and F. H. Rigler, 1975. A simple method for predicting the capacity of a lake for development based on lake trophic status. J. Fish. Res. Bd. Can., 32: 1519–1531.

Hobrough, J. E. and S. Frost, 1980. Changes in nutrient ion level of substrates and stream water due to land management in Northumberland. Environ. Pol. (Ser. A), 23: 81–93.

Likens, G. E., F. H. Bormann, N. M. Johnson, D. W. Fisher and R. S. Pierce, 1970. Effects of forest cutting and herbicide treatment on nutrient budgets in the Hubbard Brook Watershed ecosystem. Ecol. Monogr., 40: 23–47.

Likens, G. E., F. H. Bormann, R. S. Pierce, J. Eaton and N. M. Johnson, 1977. Biogeochemistry of a forested ecosystem. Springer Verlag, New York.

Marshall, B. E. and A. C. Falconer, 1973. Eutrophication of a tropical African impoundment (Lake McIlwaine, Rhodesia). Hydrobiol., 43: 109–124.

McColl, R. H. S., E. White and J. R. Waugh, 1975. Chemical run-off in catchments converted to agricultural use. N. Z. J. Sci., 18: 67–84.

Omernik, J. M., 1976. The influence of land use on stream nutrient levels. U.S. EPA Rep. No. EPA-600/3-76-014, Corvallis.

Simpson, D. E. and J. Hemens, 1978. Nutrient budget for a residential stormwater catchment in Durban, South Africa. Prog. Wat. Tech., 10: 631–643.

Simpson, D. E., V. C. Stone and J. Hemens, 1980. Water pollution aspects of stormwater run-off from a commercial land-use catchment in Pinetown, Natal. Paper presented at the Institute of Water Pollution Control (Southern African Branch) Conference, Pretoria, Republic of South Africa.

Thornton, J. A., 1980a. The Water Act, 1976, and its implications for water pollution control: case studies. trans. Zimbabwe Scient. Ass., 60: 32–40.

Thornton, J. A., 1980b. Factors influencing the distribution of reactive phosphorus in Lake McIlwaine, Zimbabwe. D.Phil. Diss., University of Zimbabwe.

Thornton, J. A. and W. K. Nduku, 1982. Nutrients in run-off from small catchments with varying land usage in Zimbabwe. Trans. Zimbabwe Scient. Ass., 61: 14–26.

Vollenweider, R. A., 1971. Scientific fundamentals of the eutrophication of lakes and flowing waters, with particular reference to nitrogen and phosphorus as factors in eutrophication. OECD Rep. No. DAS/CSI/68.27, Paris.

The solid phase: a study of pollution benchmarks on a granitic terrain
R. S. Hatherly and K. A. Viewing

The study of the pollution patterns in the Seke urban development near Salisbury resulted from the interest and the financial support of the Ministry of Natural Resources and Water Development, through the Hydrobiology Research Unit of the University of Zimbabwe (Hatherly, 1979; Hatherly and Viewing, 1981). Seke is underlain by granitic rocks of Archaean age (see

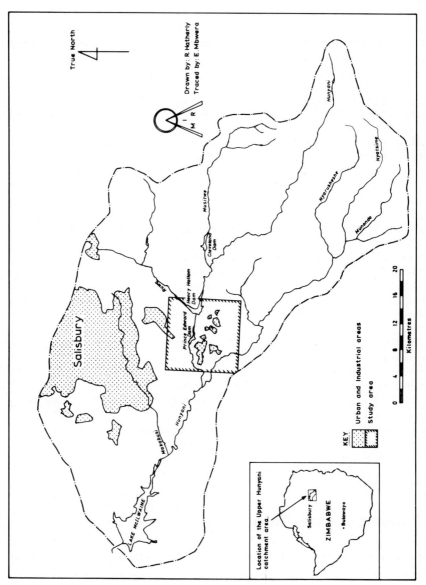

Fig. 18 The location of the Seke-Chitungwiza area in relation to the Lake McIlwaine catchment.

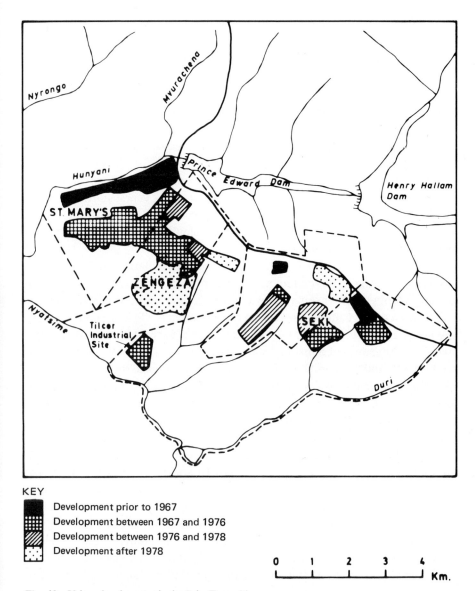

Fig. 19 Urban development in the Seke Township.

K. Munzwa, this volume) and these outcrop with sufficient frequency to obtain, by chemical analysis, an average value of the chemical composition of these rocks, as well as an indication of any significant change in composition throughout the area studied. The soils appear to be residual and relatively thin and so their chemical composition would be expected to reflect the bedrock.

In turn, stream sediments are composite samples of the products of erosion and may be compared with the geochemistry of the bedrock. Thus it is possible to determine the relationship between rock, soil and stream sediment.

The drainage sediment survey was expected to indicate the normal variation of the chemical composition of the bedrock and to reveal any anomalies that might exist. Such anomalies would be expected to result from abnormal concentrations of the chemical elements in the bedrock, and/or the effects of pollution, if these occur.

Seke is of special interest for the urban development is relatively new and it has expanded at intervals of approximately four to six years, starting with St. Mary's in 1968 and extending to Zengeza between 1968 and 1976 and to Seke in 1980 (Figs. 18 and 19). However, relatively small but well-established areas occur in each of the townships.

Thus, it may be possible to indicate the effects of pollution in a high density urban area during its development over e period of twelve years. The housing density within the townships is, on average, 25 housing units per hectare.

The area of the Chitungwiza urban development is 144 km^2 and it includes the areas designated as St. Mary's, Zengeza and Seke. It is centred approximately 20 km south of Salisbury and is bounded by commercial farming land to the north and west and by subsistence farming areas to the east and south. Drainage within the area is controlled by two main rivers: the Hunyani River which flows in a westerly direction through the northern part of the mapped area (Fig. 20) and the Nyatsime River which flows north-westwards through the south-western part of the area.

The field work was carried out using the methods of stream sediment sampling employed in the Sabi Tribal Trust Land by Topping (1974). These methods were adapted from regional geochemical mapping techniques developed in Sierra Leone (Nichol *et al.*, 1966), Northern Ireland (Webb *et al.*, 1973) and England (Webb *et al.*, 1978; Rose *et al.*, 1979). Details of the sampling and analytical methodology are given in Hatherly and Viewing (1981). Geochemical drainage survey methods have been developed mainly for mineral exploration, but during the last decade these methods were applied to the recognition of trace element deficiencies and toxicities in relation to drainage sediments, soils, vegetation and animals (Webb *et al.*, 1971; Warren and Delavault, 1971). The effects of industrial and urban pollution upon the natural variations of the geochemistry of the environment offers additional hazards which are relevant to agriculture and, ultimately, to health. In this respect, As, Pb, Cd and Hg have been confirmed as toxic to animals even in trace amounts (Underwood, 1971) and pollution levels have

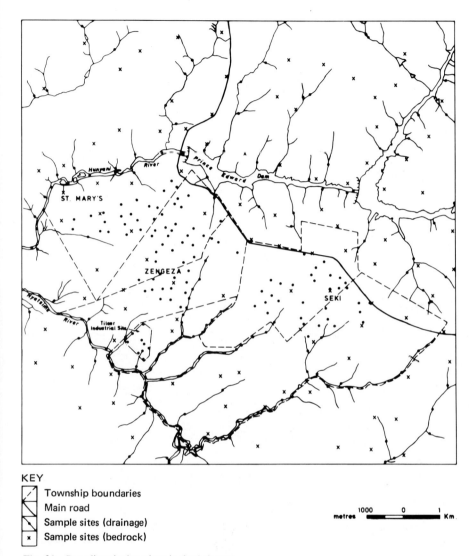

KEY
Township boundaries
Main road
Sample sites (drainage)
x Sample sites (bedrock)

Fig. 20 Sampling site locations in the Seke area.

been established for old mining and smelting activities which are monitored for present industrial wastes (Thornton, 1974). However, each potential case of chemical pollution of the environment is subject to local and perhaps to specific conditions. The geochemistry of the bedrock is a control of fundamental importance and so the granitic terrain at Seke was investigated by Hatherly and Viewing (1981) in detail.

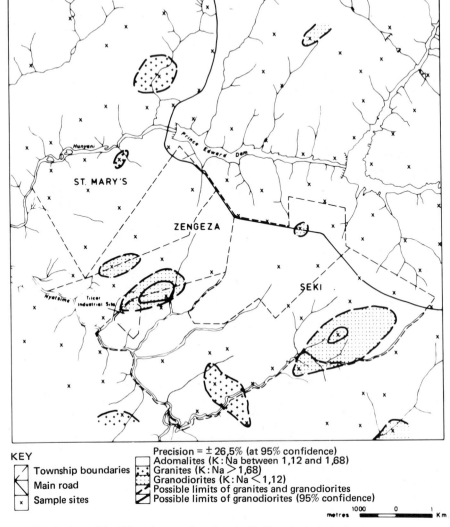

Fig. 21 Geology of the Seke urban area based on the K:Na ratios in the bedrock.

Geochemistry of the bedrock

The area covered by the Seke drainage study is underlain entirely by granitic rocks which are a part of the Salisbury adamellite (Stocklmayer *et al.*, 1978). This is confirmed by the analyses of the rocks (Fig. 21). The K:Na ratio in the bedrock over most of the area is between 1.12 and 1.68 which is in the adamellite range as proposed by Harpum (1963). The K:Na ratio also indicates the presence of several small bodies of granite with a K:Na ratio of greater than 1.68, and also of granodiorites with a ratio of less than 1.12.

The distribution of the major elements does not conform closely to the trends exhibited by the K:Na ratio, but certain features are evident (Fig. 22). Potassium (Fig. 22a) shows a relatively flat response but the content is over 3.7% within the areas designated as granites and less than 3.2% in the granodiorite areas. By contrast, the sodium content in granite areas is less than 2.5% and in the granodiorite areas is greater than 2.6%. In general, the zones of high sodium content ($> 2.57\%$) and high potassium content ($> 3.4\%$) are mutually exclusive and reflect the distribution of the rock types indicated in Fig. 21. Both lithium and rubidium follow the pattern of sodium content fairly closely. Other major elements indicate a fairly extensive zone which is relatively rich in calcium and magnesium (Fig. 22b).

The pattern of distribution of calcium, magnesium, sodium, strontium and barium (Fig. 22b) in the zone to the south-east of Prince Edward Dam surrounds and includes two occurrences of granodiorite in the south-eastern part of the study area. For this reason this pattern could be taken as an anomalous zone skirting the granodiorite. Similarly, a very small occurrence of granodiorite in the central part of the map is within a Sr-Ba anomaly. However, another granodiorite in the north-east quadrant is not associated with a Sr-Ba anomaly. It appears therefore that there is a progressive change from south to north, indicated by a gradual reduction in the amount of Ca, Mg, Ba and Sr associated with the granodiorites.

The distribution of the minor elements, Cr, Co and Ni, reflect the distribution of the major elements to a certain extent (Fig. 22c). This is evident particularly in the area south of Seke Township.

The distribution of Cu, Fe, Zn and Pb is shown in Fig. 22d. High copper contents occur mainly in the north-west of the area and across the central part of the Prince Edward Dam, and also are present in the south-west and north-east of the study area. These are isolated patches and are not related to major element distribution in the rock types present. The pattern of iron distribution corresponds fairly closely to that of copper in the west of the study area. However, in the central and eastern parts of the area this correlation is not evident. The distribution of zinc contents greater than 44 ppm closely reflects the distribution of iron. The pattern of lead distribution, characterised by areas in which the lead content exceeds 57 ppm, reflects to some extent the concentration if iron and zinc. This is evident in the anomalous area south of the Prince Edward Dam where the content reaches 97 ppm, and also in the south and south-east of the study area where the maximum content is 78 ppm. Lead is also concentrated in the bedrock in the north of the

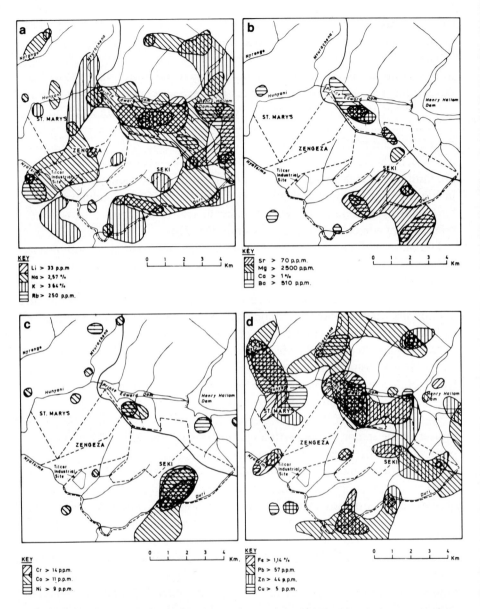

Fig. 22 Distribution of [a] lithium, sodium, potassium, rubidium, [b] strontium, magnesium, calcium, barium, [c] chromium, cobalt, nickel, [d] iron, lead, zinc and copper in the bedrock of the Seke urban area.

study area where it is associated with patchy high contents of zinc, iron, copper, nickel and cobalt.

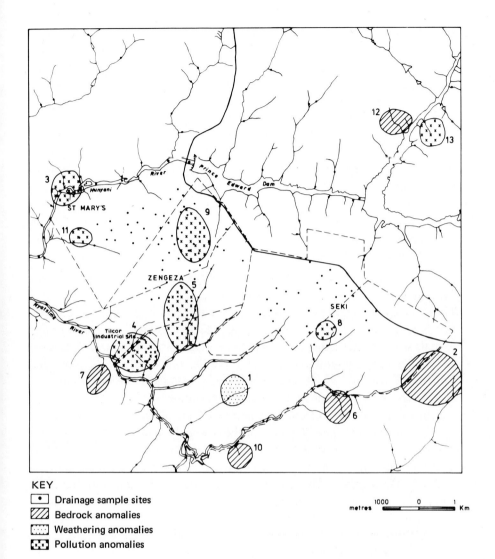

Fig. 23 Locations of the major drainage anomalies identified in the Seke Township.

In general, the distributions of the minor elements do not reflect the distributions of the different rock types shown in Fig. 21. However, Sr, Ba, Cr, Ni, and Co show a degree of correlation with the Ca-Mg rich zone described earlier. it is probably that these elements reflect regional patterns within the granites whereas Cu, Fe, Pb and Zn are more sensitive to minor differences within each rock type. The contents of Sn, W, Mo and Cd are very low in the bedrock and significant patterns are not discernable.

Description of the drainage anomalies
Thirteen areas of anomalous element concentrations were defined from the drainage reconnaissance. These areas range from single sample anomalies to areas of 10 km^2 or more in which there is generally a high content of a particular element and usually of two or more anomalous results (Fig. 23). The data for the anomalies are summarised in Table 12.

The anomalies in areas 1, 2, 6, 7, 10 and 12 are probably due to natural variations within the Seke drainage basin. The other anomalies are likely to result from artificial contamination. Contamination in areas 4, 5, 8 and 9 would appear to be caused by discarded building materials and metal scrap, whilst in areas 3, 11 and possibly also in part of 9 contamination would seem to be the result of urbanisation: e.g., automobile exhaust fumes, effluent discharges, and waste disposal. The cause of the anomaly in area 13 would indicate an artificial source but this is not obvious in the catchment.

Comparison of the metal contents of natural and artificial drainage systems with the geochemistry of the bedrock
The drainage sediments were collected from three different drainage environments and a comparison of the average element contents in these groups reveals several interesting aspects of the geochemistry of the area. The range of the elements encountered in each group is compared with bedrock data from the area in Fig. 24. This range has been calculated at the 95% confidence level as the arithmetic mean ± 2 standard deviations.

From Fig. 24 it is evident that lead and zinc show significantly higher contents in the drainage channels of the township areas than in the natural drainage. These are probably derived from contamination. The contents of lead and zinc in the natural drainages are a reasonable reflection of the bedrock. By contrast, Mn, Co, Ni and Cr appear to be concentrated in the vlei drainage, probably by a scavenging action. Lithium, sodium and strontium, which are normally constituents of fledspars, are probably dispersed chemically and so exhibit significantly lower contents in the drainage sediments than in the bedrock.

Cobalt, manganese and chromium are concentrated preferentially in the minor natural catchments and in each case the content in the sediments is significatly higher than in the bedrock. These three elements are considered to be relatively immobile in supergene silicieous environments (Rose *et al.*, 1979). Thus, they are unlikely to be transported into the major drainages but will accumulate close to their source. In addition there are higher concentrations in the minor natural drainages in the south-east of the area. There does not appear to be any accumulation of these elements from artificial sources within the township areas.

Table 12 Comparison of element contents in drainage sediments and bedrock in anomalies 1 to 13

Area	Element		Element content of anomalous stream sediment sample(s)	Background in adjacent stream sediments	Average element content in adjacent bedrock
1	Co	ppm	61	12	6
	Cu	ppm	48	8 –20	4
	Fe	%	4.54	1.4 –1.8	1.25
	Mg	ppm	3780	400 –800	2100
	Mn	ppm	4950	200 –400	240
	Ni	ppm	44	6 –21	7
2	Cr	ppm	83 –101	55	7
	Cu	ppm	34 –40	20 –37	3
	Mg	ppm	3400 –3800	1400 –3000	1550
	Ni	ppm	43 –47	21	6
3	Ba	ppm	40	320	385
	K	%	4.8	2.6 –4.3	3.56
	P	ppm	1550	90 –900	220
4	Ca	%	0.18–1.15	0.11–0.31	0.61
	Mg	ppm	740 –6000	800	1500
	Ni	ppm	12 –46	6 –21	6
	Pb	ppm	45 –156	16 –54	50
	P	ppm	524 –1857	90 –900	130
	pH		6.2 –8.9	6.8 –7.8	
	Ti	ppm	2500 –8700	2000 –8000	660
	Zn	ppm	31 –320	30 –56	35
5	Al	%	6.7 –11.7	6.2 –7.8	8.2
	Cu	ppm	9 –120	8 –20	2
	CxCu	ppm	2.1 –54.2	2.5 –6.1	0.2
	K	%	2.91–4.90	2.6 –4.3	3.1
	Pb	ppm	73 –81	54	50
6	Ca	%	1.55		0.89
	Mg	%	1.6		0.24
	Ni	ppm	26		10
	Sr	ppm	55		81
7	Cu	ppm	66		8
	CxCu	ppm	25.6		2.9
8	Ba	ppm	540		
	Ca	%	1.13		
9	Al	%	11.7		7.8
	Pb	ppm	139		42
	Zn	ppm	153		35
10	Ba	ppm	502		
	Ti	ppm	10680		
	P	ppm	1722		
11	Pb	ppm	126		42
	K	%	5.30		3.54
	Ti	%	1.07		0.05
12	Al	%	10.9		
13	Fe	%	6.2		

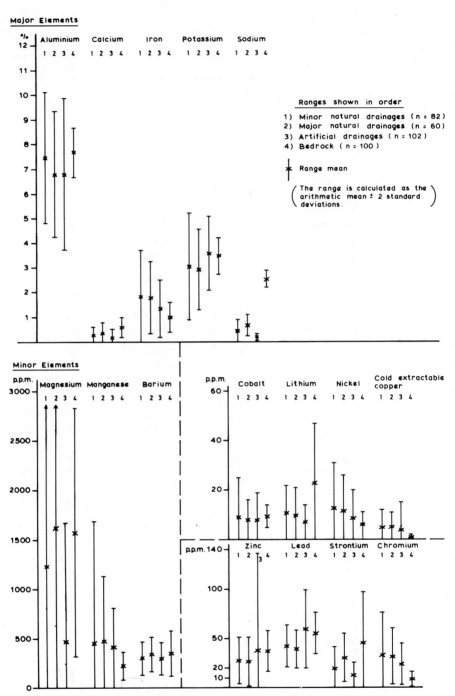

Fig. 24 Elemental distributions in natural and artificial drainages compared with those in the bedrock.

In contrast, Ba, Ca and Sr are slightly enriched in the larger natural drainage channels but the contents in the sediments are lower than those in the bedrock. These elements are present normally in feldspars. On entering the drainage system they are probably mobile and are likely to be chemically dispersed.

Geochemistry of the soils

The soils are relatively thin, particularly in the watershed areas, and the weathered bedrock in places is recognisable at depths of less than 1 m. A prominent pebble horizon consisting of rounded quartz pebbles 1 to 2 cm in size is present in some parts of the area. In general this pebble horizon occurs at a depth of 0.5 to 1.5 m but it is not consistent over the whole area. The material overlying this horizon is believed to be transported, but sorting of particle sizes is slight and a large proportion of the grains are angular or sub-rounded in shape. Thus the transport of material from nearby is indicated.

In the background areas, a significant change in the geochemistry of the soils occurs near the pebble horizon, and in areas where the pebble horizon is not developed at the base of the A soil horizon. Iron, lead, zinc, nickel, copper, cobalt, and magnesium all show higher contents within and below the pebble horizon, whereas sodium and manganese exhibit trends from the top of the pebble horizon to approximately 20 cm below. This is related to the transition from the A to B soil horizons and the pebble horizon is in places coincidental with this transition.

Two anomalies were investigated to illustrate the sources of anomalous amounts of metals in the drainage systems. Anomaly 2 is characterised by high contents of Ni with a contrast of greater than twice the background (Fig. 25a). Although the bedrock in the anomalous pit was not exposed, it is clear from the analytical results of the drainage, pit and rock samples that the anomaly reflects the presence of granodiorite and probably of weathered-mafic xenoliths in the vicinity (Figs. 21 and 22c). Thus the nickel anomaly reflects the chemical composition of the bedrock and it is not derived from contamination or sources of pollution.

By contrast, Pb is concentrated in anomaly 5 and is believed to result from contamination within the township area (Fig. 25b). The ratios of part extractable to total extractable lead in the anomalous pit ranged from 0.18 to 0.35 as compared to values from 0.09 to 0.19 in the background pit. The highest ratio was from the A soil horizon, above the pebble horizon, in the anomalous pit and this indicates the readily soluble form of lead which is present. These results indicate that the anomalous lead contents of the drainage samples

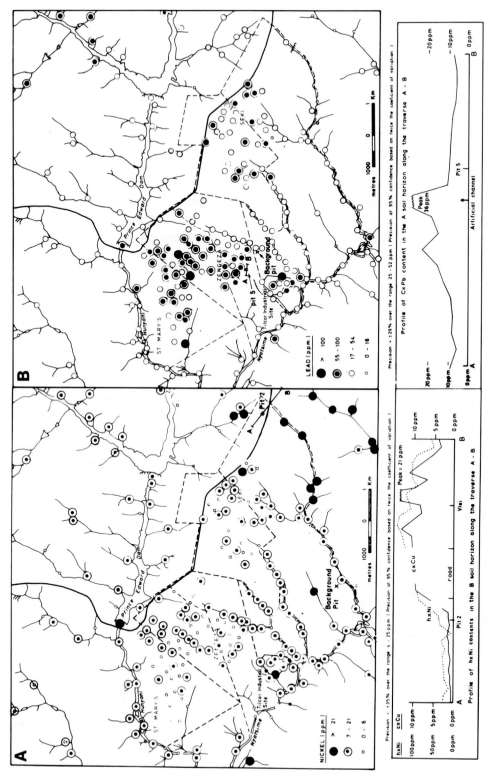

Fig. 25 The distribution of [A] nickel and [B] lead in the Seke urban area.

from the urban area result from sources of pollution, and almost certainly from motor vehicle exhaust systems.

In general, the chemical similarity between the rocks, soils and sediment is close, but variations exist which indicate erosion from different soil horizons within the area. Thus for some elements, such as Mn, Mg, Ni and Cu, the drainage sediments reflect the composition of the bedrock and for others, such as K, Na, Fe and Al, they do not due, probably, to leaching or illuviation within the soil profile, to selective gravity concentration, and to scavenging effects within the drainage sediments. In general, the soils above the pebble horizon reflect the geochemistry of the bedrock closely. However, the presence of the pebble horizon itself indicates that the soils above are transported, particularly if the pebbles are sub-anguar or rounded. The soils at Seke contain a large proportion of angular material and an examination by scanning electron microscope of the surface texture of a number of grains from the A soil horizon indicated that some chemical activity is evident but that mechanical weathering is predominant. Thus it is believed that the movement of the heavy metals into the drainage system is accomplished mainly by physical transport. A relatively small proportion is likely to be transported in solution and to be concentrated by Fe, Mn and organic matter in the vlei areas.

Artificial drainage channels were sampled by Hatherly and Viewing (1981) in order to recognise sources of pollution; if these were found to exist. Due to the regular, and in most cases rectilinear, pattern of the drainages it was difficult to assess the catchment area affecting each sample site and these were examined individually. Unlike normal weathering processes which gradually release a proportion of the relevant elements into the drainage, artificial channels within the urban areas may be subject to transient anomalies. For example, a relatively uncontaminated site may be subjected to high metal burdens by the dumping of debris close to or in the channel upstream of the sampling site. The site may return quickly to its original uncontaminated status following the removal, by flushing, of the polluting material.

Despite these possibilities, patterns of high metal concentration associated with urban development are evident in the Seke area and there is believed to be a concentration of heavy metals related to time. The distribution of lead illustrates this trend (Table 13).

A correlation exists between the age of the urban areas and the average lead content in the artificial drainages. The unpolluted natural drainage in the Seke area contain an average of 40 ppm of lead but may contain as much as 60 ppm. There appears to be a slight increase of lead, to about 50 ppm, during the initial development followed by a slow increase in concentration at a rate of about 1 ppm yr^{-1}. Concentrations above 40 ppm are likely to be due to

Table 13 Lead content of sediments in artificial drainages

Area	Approx. age (years)	No. of samples	Range in sediments (ppm)	Mean [Pb] drainage sediments (ppm)		Ratio cx/Lx	Range in bedrock (ppm)
				Lx	cx		
Seke (older part)	15	5	32–78	61	14.8	0.24	54–67
St. Mary's	12	21	42–126	59	13.7	0.23	27–50
Zengeza (older part)	10	25	45–139	68	21.8	0.32	42–57
Tilcor Industrial Area	8	9	46–156	68	21.9	0.32	47–53
Zengeza (newer part)	5	15	32–81	57	13.1	0.23	50–69
Seke (newer part)	1	27	23–90	50	15.8	0.32	48–54
Seke background				40			50–97

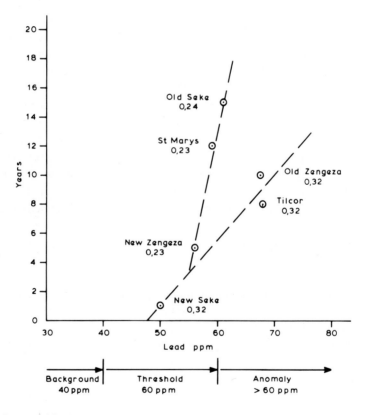

Fig. 26 Identification of threshold levels of lead pollution in the Seke area showing increasing concentrations of lead with age of development.

contamination and a period of about 10 years is necessary for the artificial drainages to reach an average level of 60 ppm.

The mean contents of cxPb and the ratio of cx:total Pb indicate that different levels of pollution exist within the various townships. In the older part of Seke, St. Mary's and the newer part of Zengeza this ratio is 0.23 to 0.24 whereas in the older part of Zengeza, Tilcor Industrial Area and the newer parts of Seke the ratio is 0.32. This indicates that the contribution to the drainage system of lead from polluting sources is greater in the latter three areas, probably due to a relatively high density of traffic and to the presence of light industries (Fig. 26).

The chemical compositions of granitic and other rocks are known to vary significantly. For this reason the selection of background and threshold values for the Seke area is unlikely to be suited to other zones of urban and industrial development in Zimbabwe and elsewhere. Thus the threshold value, or the upper limit of normal chemical variation in the sample media, or the pollution benchmark (to offer three alternative terms of definition) requires to be established for each area of interest and potential development. With this in mind, the threshold values for the Seke area are shown in Table 14 based on the analyses of stream sediments in the minus 150 μm size class.

Table 14 Mean threshold levels for heavy metals in the Seke urban area

Element	Threshold	Element	Threshold
Copper	30 ppm	Chromium	30 ppm
cx Copper	12 ppm	Cobalt	9 ppm
Lead	60 ppm	Manganese	450 ppm
cx Lead	14 ppm	Nickel	12 ppm
Zinc	50 ppm	Titanium	3700 ppm
Iron	2.0 %		

References

Harpum, J. R., 1963. Petrographic classification of granitic rocks in Tanganyika by partial chemical analysis. Records of the Geological Survey of Tanganyika, 10: 10–88.

Hatherly, R. S., 1979. Multi element drainage reconnaissance in the Seki Township. In: K. A. Viewing, Tenth annual report. Inst. Min. Research, University of Rhodesia, Rep. No. 33: 36A–38.

Hatherly, R. S. and K. A. Viewing, 1981. The geochemistry of the country around Seke Township, Salisbury – A study of threshold levels or pollution benchmarks on a granitic terrain. Inst. Min. Research Rep. to the Ministry of Natural Resources and Water Development.

Nichol, I., L. D. James and K. A. Viewing, 1966. Regional geochemical reconnaissance in Sierra Leone. Trans. Inst. Min. Metall., 75 (B): 147–161.

Rose, A. W., H. E. Hawkes and J. S. Webb, 1979. Geochemistry in mineral exploration. Academic Press, London.
Stagman, J. G. (Director), 1977. Provisional geological map of Rhodesia. Revised edition. Rhodesian Geological Survey, Salisbury.
Stocklmayer, V. R. *et al.*, 1978. The geology of the Salisbury Enterprise gold belt. Annals of the Rhodesian Geological Survey, 4: 1–12.
Thornton, I., 1974. Applied geochemistry in relation to mining and the environment. Proc. Inst. Mining Metall. Symp. Minerals and the environment – The effect of the mineral industry in environmental quality and mineral resources. pp. 1–16.
Topping, N. J., 1974. Regional geochemical drainage reconnaissance in a granitic terrain – An interim report. Inst. Min. Research, University of Rhodesia, Rep. No. C56.
Underwood, E. J., 1971. Trace elements in human and animal nutrition. New York Academics, New York.
Warren, H. V. and R. E. Delavault, 1971. Variations in the copper, zinc, lead and molybdenum contents of some vegetables and their supporting soils. In: H. L. Cannon and H. C. Hopps, Environmental geochemistry in health and disease. The Geological Soc. of America, Memoir 123.
Webb, J. S., I. Thornton and I. Nichol, 1971. The agricultural significance of regional geochemical reconnaissance in the United Kingdom. In: Anon., Trace elements in soils and crops. Tech. Bull. No. 21.
Webb, J. S. *et al.*, 1973. Provisional geochemical atlas of Northern Ireland. Applied Geochem. Research Group, Imperial College of Science and Technology, London, Tech. Comm. No. 60.
Webb, J. S. *et al.*, 1978. The Wolfson geochemical atlas of England and Wales. Applied Geochem. Research Group, Imperial College of Science and Technology, London. Clarendon Press, Oxford.

Insecticides in Lake McIlwaine, Zimbabwe
Yvonne A. Greichus

The organo-chlorine insecticides DDT, dieldrin and aldrin are still in common usage in much of Africa. These chemicals are used in the control of insect-borne diseases and in the treatment of crops. However, little information is available regarding their distribution.

As a result of the deleterious environmental effects of such insecticides their use has been restricted in North America and Europe. In view of the paucity of knowledge on the presence and persistence of insecticides in the African environment, some preliminary surveys were performed in the early 1970s.

Billing and Phelps (1972) determined insecticide residue levels in endogenous animals from a variety of areas. Their results indicated that highest values of insecticide residues were obtained in areas where agricultural land use was best developed. In the Lake McIlwaine catchment area, total DDT concentrations in a Black Flycatcher *(Melaenornis pammelaina)* were nearly twice the next highest concentration obtained in a game reserve.

Table 15 Organo-chlorine insecticide residues in birds in the Lake McIlwaine catchment in $\mu g\,g^{-1}$ dry weight (ppm); nd = not detected

Species	Sample	No. samples	Date	DDT	DDD	DDE	Total DDT	Dieldrin
Black Flycatcher[a] *(Melaenornis pammelaina)*	liver	2	1965	33.1	4.1	79.8	117.0	nd
Black-headed Heron[c] *(Ardea melanocephala)*	egg	1	1972	nd	nd	12.7	12.7	nd
Egyptian Goose[c] *(Alapochen aegyptiacus)*	egg	1	1972	nd	nd	nd	nd	nd
Fiscal Shrike[c] *(Lanius collaris)*	chick	1	1972	nd	nd	13.7	13.7	nd
Masked Weaver[c] *(Ploceus velatus)*	eggs + chicks	4+2	1972	0.3	0.8	11.6	12.7	nd
Red Bishop[c] *(Euplectes orix)*	eggs	3	1973	0.5	nd	7.8	8.3	nd
White-breasted Cormorant[b] *(Phalacrocorax carbo lucidus)*	carcass	10	1974	0.27	0.81	11.0	12.1	0.92
	brain	10	1974	0.15	0.06	2.5	2.7	1.4
	feathers	10	1974	0.06	<0.01	0.41	0.48	0.09

[a] Billings and Phelps (1972).
[b] Greichus et al. (1974).
[c] Whitwell et al. (1974).

Further surveys undertaken by Whitwell *et al.* (1974) supported the conclusions drawn by Billing and Phelps (1972). Their data also suggested that urban areas may be a source of insecticides to the environment because insecticides are widely used in the control of garden pests. Total DDT concentrations in both urban and agricultural areas exceeded those found in animals from other habitats by an order of magnitude. In addition, Whitwell *et al.* (1974) noted that pesticides tend to accumulate in aquatic habitats, particularly in urban and agricultural areas, and recommended further monitoring. Selected data on residue levels found in birds and eggs within the Lake McIlwaine catchment area are presented in Table 15. Whitwell *et al.* (1974) found no evidence of a dangerous accumulation of residues in the animals they examined.

Thus, in light of the evidence supplied by these earlier studies, Lake McIlwaine seemed an ideal place to conduct further studies. Lake McIlwaine is situated approximately 35 km south-west of Salisbury, the capital of Zimbabwe. The city is located in the northern portion of the lake catchment area which consists mainly of agricultural land and natural woodland (see K. Munzwa, this volume). Sewage effluent produced by the city is discharged into the lake via the Makabusi and Marimba Rivers after treatment.

Studies conducted during 1974 by Greichus *et al.* (1978a) revealed the following results: DDT, DDD, DDE and dieldrin were present in the Lake McIlwaine ecosystem, whilst aldrin, endosulphan and several other hydrocarbon pesticides were not detected (Table 16). The ecological magnification of insecticide residues within the food chain was demonstrated. Total average insecticide residues increased in concentration with increasing trophic level as indicated by the concentration in water (< 0.0002 ppm), bottom sediments and plankton (0.06 ppm), aquatic insects (0.35 ppm), large bream fish (0.57 ppm), and in the White-breasted Cormorant *(Phalacrocorax carbo lucidus)* at the top of the food chain (13 ppm).

Studies conducted by Greichus *et al.* (1973, 1977, 1978a) on Hartbeespoort Dam and Voëlvlei Dam in the Republic of South Africa and on Lake Poinsett, USA also demonstrated ecological magnification (Table 17). The ecological magnification of insecticides as expressed by the ratio of fish:water residues was 950×, 1600×, 1650× and 3300× for Poinsett, Voëlvlei, McIlwaine, and Hartbeespoort, respectively. Assuming that cormorants and darters do not consume large fish, the bioaccumulation ratios of insecticide residues of fish:bird are 29×, 31×, 39× and 166× for Hartbeespoort, Voëlvlei, McIlwaine and Poinsett, respectively. The higher bioaccumulation factor for Lake Poinsett may be due to higher concentrations of insecticide residues in the fish consumed by the cormorants in their wintering areas. Young cormorants

Table 16 Average concentration of insecticides in the Lake McIlwaine ecosystem in $\mu g\, g^{-1}$ dry weight except water (after Greichus *et al.* 1978a)

Description	No. samples	DDE	Dieldrin	DDD	DDT	Total insecticides
Water	10	0.0001	<0.0001	<0.0001	<0.0001	<0.0002
Bottom sediments	10	0.015	0.004	0.040	0.002	0.061
Plankton	5[a]	0.01	<0.01	0.02	0.02	0.06
Oligochaetes	1[a]	0.18	0.08	0.33	0.14	0.73
Benthic insects	1[a]	0.13	<0.01	0.12	0.11	0.36
Fish						
Dwarf Bream	5[b]	0.12	0.07	0.16	0.04	0.39
Spot Tail	5[b]	0.24	0.04	0.10	<0.01	0.38
Greenheaded Bream	5[b]	0.08	0.03	0.10	<0.01	0.22
Greenheaded Bream	5[c]	0.13	0.12	0.18	0.14	0.57
Bird						
Cormorant carcass	10	11.0	0.92	0.81	0.27	13.0

[a] Each sample consists of a composite collected from all over the lake.
[b] Each sample consists of a composite of 10 fish ranging in weight from 6 to 40 g.
[c] Individual fish ranging in weight from 578 to 824 g.

Table 17 A comparison of total organo-chlorine insecticide residues in Lake McIlwaine with those from other African lakes and a North American lake in $\mu g\, g^{-1}$ dry weight; nr = not reported

Lake	Location	Water	Sediment	Plankton	Benthos	Fish	Birds
McIlwaine[a]	Zimbabwe	<0.0002	0.061	0.06	0.36	0.39	13.0
Kariba[b,c,d]	Zimbabwe	nr	nr	nr	nr	9.38[e]	6.2
Hartbeespoort[a]	South Africa	0.0003	0.045	0.59	nr	1.0	25.5
Voëlvlei[a]	South Africa	<0.0002	0.013	nr	0.29	1.16	10.0
Nakuru[a]	Kenya	<0.0002	<0.002	0.12	0.06	0.05	nr
Tanganyika[f]	Burundi	nr	nr	nr	nr	1.3	nr
Poinsett[a]	USA	0.0002	0.002	nr	nr	0.19	31.0

[a] Greichus *et al.* (1973, 1977, 1978a, 1978b).
[b] Billings and Phelps (1972).
[c] Whitwell *et al.* (1974).
[d] Wessels *et al.* (1980).
[e] Crocodile eggs.
[f] Deelstra *et al.* (1976).

feeding exclusively on Lake Poinsett fish had much lower insecticide residue levels (1.5 ppm dry weight) than the adults (29.0 ppm dry weight) (Greichus *et al.*, 1973).

Sun-dried fish taken from the northern end of Lake Tanganyika had total residues of DDE + DDD + DDT varying from 0.45 ppm to 2.39 ppm with an overall average of 1.3 ppm (Deelstra et al., 1976). Small amounts of dieldrin, endrin and lindane were also found in some fish. The levels of total insecticides in fish in the northern end of Lake Tanganyika appear to be greater than those in Lake McIlwaine fish.

The most recent surveys of pesticide residues in the Lake McIlwaine system have been undertaken by the Zimbabwe Government as part of a national pesticide monitoring programme. Table 18 presents some preliminary results from this programme (Working Party on the Monitoring of Organo-chlorine Pesticides in the Environment, 1979) in comparison with the data reported by Greichus et al. (1978a). These data indicate that total DDT concentrations in at least two species of fish, the Dwarf and Greenheaded Bream (*Haplochromis darlingi* and *Sarotherodon macrochir,* respectively), have increased substantially between 1974 and 1979. Differences in experimental technique may account for some of the observed discrepancy. Nevertheless, it is a disturbing trend and will bear watching in the future.

The degree of urban and agricultural development in lake catchments can be seen to play a major role in determining the amounts of insecticides in the

Table 18 Insecticide residues found in commercially important fish species in Lake McIlwaine between 1974[a] and 1979[b] in $\mu g\ g^{-1}$ dry weight

Species	No. samples	Approx. mass (g)	Year	DDT	DDD	DDE	Total DDT	Dieldrin
Dwarf Bream								
(*Haplochromis darlingi*)	10[c]	23–40	1974	0.04	0.16	0.12	0.32	0.07
	2[c]	25	1979	0.29	0.33	0.58	1.20	0.10
Spot Tail								
(*Alestes imberi*)	10[c]	6–9	1974	<0.01	0.10	0.24	0.34	0.04
Greenheaded Bream								
(*Sarotherodon*	10[c]	23–30	1974	<0.01	0.10	0.08	0.18	0.03
macrochir)	10	578–824	1974	0.14	0.18	0.13	0.45	0.12
	2[c]	650–700	1979	0.22	0.36	0.69	1.27	0.10
Sharptooth Catfish								
(*Clarias gariepinus*)	1[c]	700	1979	0.26	0.25	1.00	1.51	0.99

[a] Greichus et al. (1978a).
[b] Working Party on the Monitoring of Organo-chlorine Pesticides in the Environment (1979).
[c] Composite samples.

environment. Hartbeespoort Dam receives pollutants from large urban and industrial complexes and from agricultural sources, and thus has higher levels of insecticide residues than Lake McIlwaine. The insecticide residues in Lake McIlwaine also arise from agricultural and urban sources but the contribution from urban sources is not as great as those to Hartbeespoort Dam. Lake Nakuru, Voëlvlei Dam and Lake Poinsett are comparatively unaffected by urban pollution and in general have lower levels of insecticide residues than Lake McIlwaine. Wessels *et al.* (1980) determined that residue levels were low in crocodile egges collected from areas remote from human habitation at Lake Kariba, Zimbabwe/Zambia. However, increased levels were found in eggs from localities adjacent to human populations or near rivers draining agricultural areas. It is difficult, however, to assess the relative importance of agricultural sources of insecticides without more information on the size and agricultural practices within each lake catchment area.

References

Billing, K. J. and R. J. Phelps, 1972. Records of chlorinated hydrocarbon pesticide levels from animals in Rhodesia. Trans. Rhod. Scient. Ass., 55: 6–9.
Deelstra, H., J. L. Power and C. T. Kenner, 1976. Chlorinated hydrocarbon residues in the fish of Lake Tanganyika. Bull. Environ. Contamin. Toxicol., 15: 689–698.
Greichus, Y. A., A. Greichus and R. J. Emerick, 1973. Insecticides, polychlorinated biphenyls and mercury in wild cormorants, pelicans, their eggs, food and environment. Bull. Environ. Contamin. Toxicol., 9: 321–328.
Greichus, Y. A., A. Greichus, B. D. Amman, D. J. Call, D. C. D. Hamman and R. M. Pott, 1977. Insecticides, polychlorinated biphenyls and metals in African lake ecosystems. I. Hartbeespoort Dam, Transvaal and Voëlvlei Dam, Cape Province, Republic of South Africa. Arch. Environ. Contamin. Toxicol., 6: 371–383.
Greichus, Y. A., A. Greichus, H. A. Draayer and B. E. Marshall, 1978a. Insecticides, polychlorinated biphenyls and metals in African lake ecosystems. II. Lake McIlwaine, Rhodesia. Bull. Environ. Contamin. Toxicol., 19: 444–453.
Greichus, Y. A., A. Greichus, B. D. Amman and J. Hopcroft, 1978b. Insecticides, polychlorinated biphenyls and metals in African lake ecosystems. III. Lake Nakuru, Kenya. Bull. Environ. Contamin. Toxicol., 19: 454–461.
Wessels, C. L., J. Tannock, D. Blake and R. J. Phelps, 1980. Chlorinated hydrocarbon insecticide residues in *Crocodilus niloticus* Laurenti eggs from Lake Kariba. Trans. Zimbabwe Scient. Ass., 60: 11–17.
Whitwell, A. C., R. J. Phelps and W. R. Thomson, 1974. Further records of chlorinated hydrocarbon residues in Rhodesia. Arnoldia Rhod., 6: 1–8.
Working Party on the Monitoring of Organo-chlorine Pesticides in the Environment, 1979. Unpublished report. Department of Research and Specialist Services, Ministry of Agriculture, Salisbury.

5   Biology

An SEM study of bacteria and zooplankton food sources in Lake McIlwaine
Monika Boye-Chisholm and R. D. Robarts

A lake is an assemblage of a very large number of living organisms and non-living matter which show complex inter-relationships and interactions. In the present volume data are presented on various aspects of the abiotic and biotic components of Lake McIlwaine. The present study was an initial attempt to examine some of the inter-relationships between the various components of the lake using a scanning electron microscope (SEM). While most components of Lake McIlwaine have to some degree been the object of study, the bacteria (with the exception of the blue-green bacteria/algae) have been overlooked. Our study examines some aspects of these important organisms in Lake McIlwaine especially as a food source for zooplankton. The procedure of Paerl and Shimp (1973) was used in the preparation of algae, bacteria and detritus for examination with the SEM; unfortunately membrane filters were not available at the time of this study (1976) and glass fibre filters were used to mount the material. Details of methods are given in Boye (1976).

Figures 1 to 5 are representative of the photomicrographs obtained. Types of bacterial attachment to the substrate are shown in Figs. 1 and 2. Fig. 1 shows fibrillar appendages formed by bacilli. Coccoid bacteria also produce similar appendages with which, in Fig. 2, they are secured to a flagellate. In Fig. 1 bacterial attachment by a long filament can also be seen.

Scanning electron microscopy has shown a number of means of bacterial attachment to suspended matter. Among these are adhesive stalk formations, capsular secretions, fibrillar appendages (which serve as anchors), attached webbing on which cells are located and absorption of bacterial cells to particles without the aid of cellular appendages or secretions (Paerl, 1975). In addition to the types of attachment shown in Figs. 1 and 2 we have also seen webbing attachments. A recent study of fibrillar colloids associated with algae and bacteria in lakes indicated that they promoted complex microbial associations and physical associations between cells and other suspended

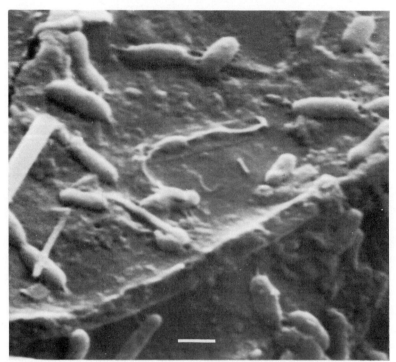

Fig. 1 Bacterial attachment to detritus in Lake McIlwaine. White bar = 1 μm.

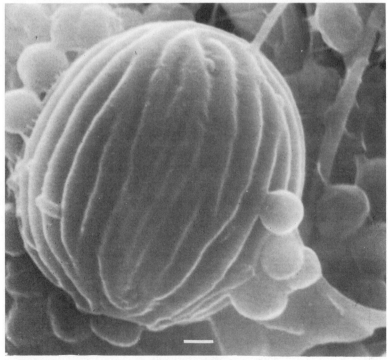

Fig. 2 Bacterial attachment to phytoplankton. White bar = 2 μm.

particles (Massalski and Leppard, 1979). The authors also noted that one cell can make more than one type of fibril at one time and that the trapping of debris and clay particles by fibril aggregates almost certainly must occur.

Results of the examination of *Tropodiaptomis* gut contents are shown in Figs. 3 to 5. The animals ingested a number of different types of algae including *Pediastrum* (Fig. 3) which is relatively rare in Lake McIlwaine, the diatom *Melosira* (Fig. 4) and *Microcystis aeruginosa,* the dominant alga in the lake (see J. A. Thornton, this volume). According to Arnold (1971) *Microcystis* is of low nutritional value to zooplankton. Further studies are required to determine if *Tropodiaptomis* in Lake McIlwaine actively ingest and digest these cells. As indicated in Figs. 3 to 5, bacteria and detritus appeared to form the major component of the zooplankton diet. *Tropodiaptomis* was the only zooplankton species dissected in our study as it was both the largest animal and the most abundant genus at the time.

Gliwicz (1969) and Kajak (1970) maintain that only very small algae (nanno-

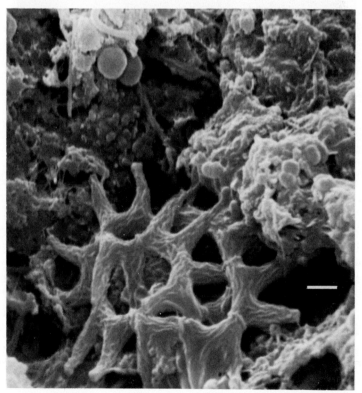

Fig. 3 Gut contents of *Tropodiaptomus* from Lake McIlwaine. White bar = 3 μm.

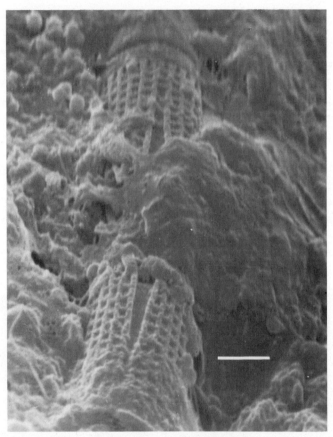

Fig. 4 *Melosira granulata* in the gut contents of *Tropodiatomus*. White bar = 10 μm.

plankton) can be utilised directly by planktonic animals. Gliwicz found that in oliogotrophic lakes the main zooplankton food components were nannoplankton followed by detritus and bacteria. In contrast, the diet of zooplankton from eutrophic lakes was 74% bacteria. Gliwicz related the change in food preference to particle size. Eutrophic lakes are characterised by net plankton usually too large to be ingested by most zooplankton. Bacteria and detritus therefore are the preferred food sources. Peterson *et al.*, (1978) have demonstrated that zooplankton grazers, such as *Daphnia,* can feed on the small natural bacterial flora of a lake.

As part of an on-going study of the zooplankton of Hartbeespoort Dam, Republic of South Africa, the gut contents of *Ceriodaphnia* and *Thermocyclops* were examined using the acridine orange and epifluorescent microscope technique of Hobbie *et al.* (1977). The preliminary results indicated that

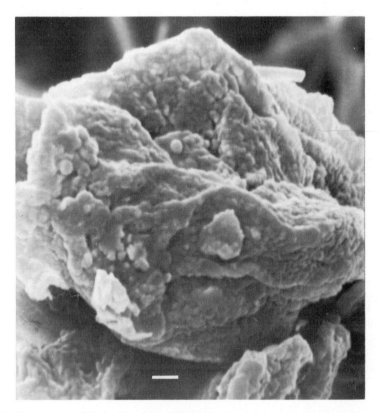

Fig. 5 Gut contents of *Tropodiaptomus*. White bar = 2 μm.

the gut contents contained several diatom genera but the majority of the food was bacteria and detritus (H. J Taussig, personal communication). Hartbeespoort Dam is eutrophic and physically similar to Lake McIlwaine, and both impoundments have phytoplankton populations dominated by *Microcystis aeruginosa*.

While the results of our study can only be considered preliminary, they indicate the role of aquatic bacteria in the formation of detrital aggregates and the probable importance of bacteria and detritus as a zooplankton food source in Lake McIlwaine. Our observations are similar to those recorded for many other lakes.

Acknowledgements

We wish to acknowledge the help of R. Çavil in preparing the SEM stubs,

T. R. C. Fernandes for operating the SEM and H. J. Taussig for allowing us to examine her zooplankton data. This work was submitted by M. Boye in partial fulfillment for the B. Sc. (Hons.) degree in the Division of Biological Sciences, University of Rhodesia and was supported by funds given to the Hydrobiology Research Unit, University of Rhodesia, by the Ministry of Natural Resources and Water Development and the City of Salisbury.

References

Arnold, E. E., 1971. Ingestion, assimilation, survival and reproduction by *Daphnia pulex* fed seven species of blue-green algae. Limnol. Oceanogr., 16: 906–920.
Boye, M., 1976. A scanning electron microscope study of the relationships between algae, bacteria, zooplankton and detritus in Lake McIlwaine, Rhodesia. B. Sc. (Hons.) thesis, University of Rhodesia.
Gliwicz, Z. M., 1969. The share of algae, bacteria and trypton in the food of the pelagic zooplankton of lakes with various trophic characteristics. Bull. Acad. pol. Sci. Cl. II Ser. Sci. biol., 17: 159–165.
Hobbie, J. E., R. J. Daley and S. Jasper, 1977. Use of Nuclepore filters for counting bacteria by fluorescence microscopy. Appl. Environ. Microbiol., 33: 1225–1228.
Kajak, Z., 1970. Some remarks of the necessities and prospects of the studies on biological production of freshwater ecosystems. Polskie Archwn. Hydrobiol., 17: 43–54.
Massalski, A. and G. G. Leppard, 1979. Morphological examination of fibrillar colloids associated with algae and bacteria in lakes. J. Fish. Res. Bd. Can., 36: 922–938.
Paerl, H. W., 1975. Microbial attachment to particles in marine and freshwater ecosystems. Microbial Ecology, 2: 71–83.
Pearl, H. W. and S. L. Shimp, 1973. Preparation of filtered plankton and detritus for study with scanning electron microscopy. Limnol. Oceanogr., 18: 802–805.
Peterson, B. J., J. E. Hobbie and J. F. Haney, 1978. *Daphnia* grazing on natural bacteria. Limnol. Oceanogr., 23: 1039–1044.

Phytoplankton, primary production and nutrient limitation
R. D. Robarts, J. A. Thornton and Colleen J. Watts

The algal community
J. A. Thornton

Species composition

The algal community of Lake McIlwaine is dominated by *Microcystis aeruginosa* Kutz for much of the year. Munro (1966) in his early study of the lake noted the predominance of this blue-green alga in the phytoplankton community. In addition, *Anabaena flos-aquae* (Lyng.) Breb. occurred frequently and Munro (1966) also reported the presence of large numbers of chlorophytes, particularly *Volvox* sp., *Eudorina* sp. and *Pediastrum* sp., as well as of the desmid, *Staurastrum* sp. Large diatoms were rarely found. More recently, however, Falconer (1973) reported that the diatom *Melosira granulata* (Ehr.) Rolfs was dominant in the impoundment during spring 1968–69 (November 1968) and that *Microcystis* sp. and *Anabaena* sp. were the dominant phytoplankters at other times of the year. This dominance of *Microcystis* and *Anabaena*, and occasionally of *Melosira*, has been reported by Mitchell and Marshall (1974), Robarts and Southall (1977) and Robarts (1979) for the period between 1968 and 1976 and has been observed since (personal observation). *Anabaenopsis tanganyikae* was also reported by Mitchell and Marshall (1974) as being amongst the dominant algae in the lake during their study of 1970–71. They also reported some *Pediastrum clathratum*. Robarts (1979) reported a large *Lyngbya contorta* population in the lake during May 1976 although *Melosira granulata* was the dominant species recorded during that month. A partial species list of the phytoplankton genera observed in Lake McIlwaine is shown in the addendum.

Seasonal variation
Robarts (1979; unpublished) has described the seasonal variation of species composition of the phytoplankton of Lake McIlwaine. *Microcystis aeruginosa* dominates the phytoplankton population of the lake for most of the year but particularly during summer (December to April). *Anabaena* and/or *Anabaenopsis* is usually the second most important alga during this period, but may predominate during other seasons, particularly spring. *Melosira* increases in importance during early winter (March to June) and may even become dominant for short periods during this season. Winter and spring (April to December) are the periods of greatest phytoplankton diversity with several species of chlorophytes and pyrrophytes being present in addition to the three dominant genera above: e.g. *Actinastrum* sp., *Scenedesmus* sp., *Chlorella* sp., *Staurastrum* sp., and *Ceratium* sp. (cf., addendum). As noted *Lyngbya* sp. was also observed in the phytoplankton during this period.

This assemblage of phytoplankton species is indicative of an enriched system. Whilst the presence and dominance of *Microcystis aeruginosa* does not in itself indicate eutrophication (as this alga can tolerate a wide range of habitat types; cf., Walmsley and Butty, 1980; Thornton and Cotterill, 1978), its dominance in association with *Anabaena* sp. and *Melosira* sp. to the virtual exclusion of other algal species is common to most eutrophic reservoirs in southern Africa for which species lists exist (Walmsley and Butty, 1980; Robarts and Southall, 1977; Osborne, 1972).

Standing crop
Whilst there has been little apparent change in the species composition of the phytoplankton in Lake McIlwaine following nutrient diversion (C. H. D. Magadza, personal communication), there has been a significant reduction in

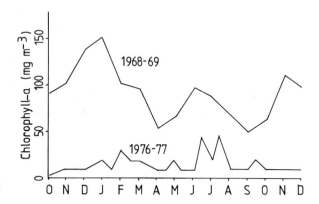

Fig. 6 Distribution of chlorophyll *a* in Lake McIlwaine surface waters at the mid-lake station during 1968–69 and 1976–77 (after Falconer, 1973; Thornton, 1980).

the surface water chlorophyll *a* concentration (Fig. 6). Maximum chlorophyll *a* concentrations of between 50 and 150 mg m^{-3} were reported by Falconer (1973) during 1968–69. Robarts (1979; unpublished) found a range in chlorophyll *a* concentrations that was similar to that observed by Falconer (1973), with surface water concentrations ranging between 12 and 140 mg m^{-3}. In a more recent study, Thornton (1980) has reported chlorophyll *a* concentrations of between 2 and 45 mg m^{-3} with a mean concentration of 15 mg m^{-3}. All three studies suggest that there are three fairly distinct growing periods during which the phytoplankton standing crop increases in size. These periods correspond to the three seasons, and would appear to be a characteristic of southern African impoundments (Walmsley *et al.*, 1978; Walmsley and Toerien, 1979). It is not clear from the available data whether these chlorophyll peaks are related to the seasonal changes in the species composition of the phytoplankton standing crop as discussed above, but such patterns are not uncommon in temperate lake systems (Fogg, 1975). Nevertheless, the data collected by Robarts (unpublished) and the data given by Munro (1966) in his Fig. 5 would seem to suggest that this is in fact the case, with *Microcystis* sp. being most abundant during summer, *Melosira* sp. during early winter, and *Anabaena* sp. during spring.

Addendum

A partial list of phytoplanktonic algae in Lake McIlwaine

1. Chlorophyta
    Volvocaceae
        Volvox sp.
        Eudorina sp.
    Hydrodictyaceae
        Pediastrum clathratum
    Oocystaceae
        Chlorella sp.
    Scenedesmaceae
        Scenedesmus sp.
        Actinastrum sp.
    Desmidaceae
        Staurastrum sp.
2. Chrysophyta
    Bacillariophyceae
        Melosira granulata
3. Cyanophyta
    Chroococcaceae
        Microcystis aeruginosa
    Nostoceae
        Anabaena flos-aquae
        Anabaenopsis tanganyikae
    Oscillatoriaceae
        Lyngbya contorta
4. Pyrrophyta
    Ceratiaceae
        Ceratium sp.

References

Falconer, A. C., 1973. The phytoplankton biology of Lake McIlwaine, Rhodesia. M. Phil. thesis, University of London.
Fogg, G. E., 1975. Algal cultures and phytoplankton ecology. University of Wisconsin Press, Madison.
Mitchell, D. S. and B. E. Marshall, 1974. Hydrobiological observations on three Rhodesian reservoirs. Freshwat. Biol., 4: 61–72.
Munro, J. L., 1966. A limnological survey of Lake McIlwaine, Rhodesia. Hydrobiol., 28: 281–308.
Osborne, P. L., 1972. A preliminary study of the phytoplankton of selected Rhodesian man-made lakes. Rhod. Sci. News, 6: 294–297.
Robarts, R. D., 1979. Underwater light penetration, chlorophyll α and primary production in a tropical African lake (Lake McIlwaine, Rhodesia). Arch. Hydrobiol, 86: 423–444.

Robarts, R. D. and G. C. Southall, 1977. Nutrient limitation of phytoplankton growth in seven tropical, man-made lakes, with special reference to Lake McIlwaine, Rhodesia. Arch. Hydrobiol., 79: 1–35.

Thornton, J. A., 1980. Factors influencing the distribution of reactive phosphorus in Lake McIlwaine, Zimbabwe. D. Phil. Diss., University of Zimbabwe.

Thornton, J. A. and N. G. Cotterill, 1978. Some hydrobiological observations on five tropical African montane impoundments. Trans. Rhod. Scient. Ass., 59: 22–29.

Walmsley, R. D. and M. Butty, 1980. Limnology of some selected South African impoundments. Water Research Commission, Pretoria.

Walmsley, R. D. and D. F. Toerien, 1979. A preliminary limnological study of Buffelspoort Dam and its catchment. J. Limnol. Soc. Sth. Afr., 5: 51–58.

Walmsley, R. D., D. F. Toerien and D. J. Steÿn, 1978. An introduction to the limnology of Roodeplaat Dam. J. Limnol. Soc. Sth. Afr., 4: 35–52.

Primary production of Lake McIlwaine
R. D. Robarts

Eutrophication is the enrichment of an aquatic ecosystem with plant nutrients (specifically phosphorus and nitrogen) resulting in an increased production at all trophic levels. In most non-eutrophic lakes phosphorus is usually the nutrient which limits algal growth (Robarts and Southall, 1977) and therefore the higher trophic levels (Melack, 1976). As effluents, with low nitrogen to phosphorus ratios, are continually added to a lake, the algal growth limiting nutrient changes from phosphorus to nitrogen. At this stage large populations of nitrogen-fixing blue-green algae may appear. If nutrient addition continues, the concentration of combined forms of nitrogen will reach levels which will inhibit nitrogen fixation but will permit other species to form dense populations (Horne, 1979). The process, if unchecked, will continue until light becomes the dominant limiting factor to the primary producers.

Eutrophic Lake McIlwaine has been the subject of considerable previous study (Falconer, 1973; Marshall and Falconer, 1973a, 1973b; Mitchell and Marshall, 1974; Munro, 1966, 1967; Nduku, 1976, Robarts, 1979; Robarts and Southall, 1975, 1977; Robarts and Ward, 1979; Thornton, 1979a, 1979b, 1980; see also this volume). Although algal bioassays of the lake indicated nitrogen was potentially the primary growth limiting nutrient, no evidence of nitrogen fixation by the blue-green algal population has been found (Robarts and Southall, 1977; A. J. Horne in Stewart, 1974). Preliminary primary productivity and light penetration data for Lake McIlwaine suggested to Robarts and Southall (1977) that light might be a more important factor regulating algal productivity than nitrogen. This was suggested by Falconer (1973) after his study of primary productivity in 1968–69. This paper is a review of the primary productivity data available for Lake McIlwaine. For specific details of the methodologies employed, readers should consult the original articles cited.

The first study of primary productivity on Lake McIlwaine was done by Falconer (1973) during 1968 and 1969. He used the oxygen light and dark bottle method and found that productivity on a unit volume basis, at optimum depth, ranged between 400 and 2000 mg O_2 m^{-3} h^{-1} (150 to 749 mg C m^{-3} h^{-1}). Falconer found that carbon fixation was usually not inhibited at the surface and that productivity generally ceased below four meters. He also gave a figure with integral hourly values but failed to give the scale. As noted above, he indicated that nutrient concentrations in Lake McIlwaine were probably not limiting to primary productivity but that light was the controlling factor.

In 1970, Mitchell and Marshall (1974) measured primary production in three Rhodesian impoundments. They also used the oxygen light and dark bottle technique. The Lake McIlwaine data for October 1970 show a depth profile with two maxima, one at the surface and the other at two metres. Maximal assimilation rates recorded were about 1580 mg O_2 m^{-3} h^{-1} (593 mg C m^{-3} h^{-1}). Below 2 m there was a rapid decline in production which Mitchell and Marshall concluded was due to the sharp decline of light as indicated by the Secchi disc value of 1.02 m. At 8 m primary production was 100 mg O_2 m^{-3} h^{-1} (38 mg C m^{-3} h^{-1}). No measurements were made below this depth.

Robarts and Southall (1977) measured Lake McIlwaine primary production in February 1975. The work was done at the same station as that of Falconer (1973) and Mitchell and Marshall (1974), station SM-4. The carbon-14 (^{14}C) light and dark bottle technique was used. During all of the above studies the dominant alga was *Microcystis aeruginosa*. Robarts and Southall's data showed that primary productivity at the surface was depressed so that maximum productivity occurred at the level where light penetration (450 to 1150 mμ) was reduced to 25% of the surface value. This depth was 0.5 m and a carbon fixation rate of 263 mg C m^{-3} h^{-1} (701 mg O_2 m^{-3} h^{-1}) was measured. The integrated value for the water column was 351 mg C m^{-2} h^{-1} (936 mg O_2 m^{-2} h^{-1}). Below 0.5 m primary productivity was quickly reduced so that by 3 m only 4 mg C m^{-3} h^{-1} (11 mg O_2 m^{-3} h^{-1}) were being fixed.

Robarts and Southall (1977) presented a semi-log plot of light penetration and carbon assimilation. The similarity between the curves suggested to them a light limitation of production in the euphotic zone of Lake McIlwaine. The nature of the Lake McIlwaine production profile was similar to the Type 1 profiles obtained by Findenegg (1964) for eutrophic lakes that have an abundance of nutrients and a high standing stock of phytoplankton that self-shade and reduce light penetration.

In order to reduce the algal population of Lake McIlwaine, the City of Salisbury implemented a programme of sewage diversion to land (see J. McKendrick, this volume). In 1974–75 approximately 50% of the sewage effluent was diverted and from June 1975 it was all placed on land. In June 1975, Robarts (1979) began an extensive study of underwater light penetration, chlorophyll a and primary production of Lake McIlwaine to determine the regulating effect of underwater light conditions had on algal primary productivity. The study was also undertaken to ascertain if any significant changes in algal productivity had occurred because of the diversion programme.

Primary production was measured with the ^{14}C light and dark bottle technique at the central lake station, SM-4, in Lake McIlwaine as was used in previous studies. Sampling was fortnightly from June 1975 to May 1976 inclusive. The data were analysed using Talling's (1965) model for phytoplankton photosynthesis. During the study *Microcystis aeruginosa* was usually the dominant alga followed by *Anabaena/Anabaenopsis*. In April and May 1976 *Melosira* was the dominant alga.

The vertical extinction coefficients (ε) indicated that blue light was rapidly attenuated and that green light usually penetrated the furthest. Secchi disc transparency ranged between 0.61 m and 1.6 m. The depth of the euphotic zone $Z_{eu} = 3.7 / \varepsilon_{min}$, ranged between 1.3 and 3.6 m and was inversely correlated with algal standing crop (12 to 95 mg m^{-3} chlorophyll a).

The relationship between the minimum vertical extinction coefficient (ε_{min}) and the mean concentration of chlorophyll a in the upper 2 m of the water column is expressed by the regression equation (Robarts, 1979):

$$\varepsilon_{min} = 0.0207 B + 0.76 \qquad (r = 0.89) \tag{1}$$

The slope of the regression equation is ε_s and indicates the increment of ε_{min} per unit increment of algal concentration [(mg m^{-2})$^{-1}$]. The parameter is a useful measure of the self-shading properties of the phytoplankton. The importance of a relatively high ε_s value, as recorded in Lake McIlwaine, is that it will not allow large euphotic zone algal populations to develop and therefore reduces unit volume primary productivity.

Chlorophyll accounted for a minimum of 26.6% of ε_{min} and a maximum of 73.9% with a mean value of 52.9% (Robarts, 1979). The value of 0.76 in the above regression equation indicated that the proportion of light extinction due to factors other than chlorophyll was low. The ε_s value of 0.0207 and the low background extinction value of 0.76 indicated that the algal population of Lake McIlwaine exhibited a strong self-shading potential and could account for the relatively small algal population recorded as compared with Lake

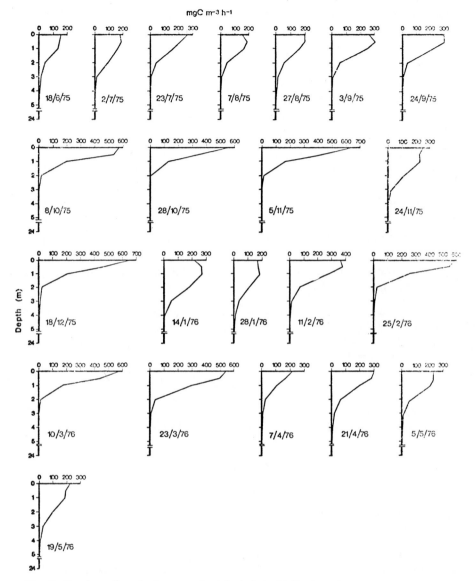

Fig. 7 Depth profiles of primary productivity in Lake McIlwaine (after Robarts, 1979).

George (Uganda) and Loch Leven (Scotland) (Robarts, 1979).

The photosynthesis-depth profiles recorded for Lake McIlwaine are shown in Fig. 7. Depressed rates of photosynthesis at the surface occurred on a few occasions. Lake McIlwaine is characterised by a shallow zone, usually not exceeding 3 m, here photosynthesis occurs. The maximum rate of carbon incorporation occurred between the surface and 0.5 m and decreased rapidly thereafter. The maximum rate of photosynthesis, on a unit volume basis, of 653 mg C m^{-3} h^{-1} (1742 mg O$_2$ m^{-3} h^{-1}) was recorded at the surface. The integrals, determined by planimetry, of the depth profiles in Fig. 7 ranged between 248 and 635 mg C^{-2} h^{-1} (661 to 1694 mg O$_2$ m^{-2} h^{-1}). On a daily basis, productivity ranged between 1.64 and 6.03 g C m^{-2} d^{-1} (4.37 to 16.08 g O$_2$ m^{-2} d^{-1}). Robarts (1979) calculated an annual primary production rate of 1.43 kg C m^{-2} (3.81 kg O$_2$ m^{-2}) for Lake McIlwaine. The lake is the second most productive yet studied in southern Africa (Table 1). These data, when compared with previous data from other studies, indicated that the sewage diversion programme had not yet had a significant effect on primary production.

Robarts (1979) examined various factors which may have had a regulating role on primary production. He found that the ratio $A_{max}/\varepsilon_{min}$ (light saturated rate of net photosynthesis per unit volume/minimum vertical extinction coefficient) accounted for most (68%) of the variation in the hourly integral rate of photosynthesis. The measurement of algal standing crop as chlorophyll a

Table 1 A comparison of Lake McIlwaine primary production with that recorded for other warm, freshwater lakes. Oxygen production data converted using the relationship: mg C = 0.375 mg O$_2$

Lake	A_{max} (mg C m^{-3} h^{-1})	ΣA (mg C m^{-2} h^{-1})	$\Sigma\Sigma A$ (g C m^{-2} d^{-1})	Reference
Castanho (Brazil)	–	–	0.50–1.50	Schmidt (1973)
Chad (Chad)	66–336	61–318	0.70–2.69	Lemoalle (1973)
Crescent Is. Crater (Kenya)	19–68	105–293	1.13–3.15	Melack (1979)
George (Uganda)	–	375–750	1.95–5.80	Ganf (1975)
Hartbeespoort (South Africa)	12–5916	47–3381	0.40–30.9	Robarts (in prep.)
Kinneret (Israel)	–	–	0.56–8.05	Berman (1976)
Lanao (Philippines)	–	–	0.40–5.00	Lewis (1974)
McIlwaine (Zimbabwe)	155–653	248–653	1.64–6.03	Robarts (1979)
Naivasha (Kenya)	56–90	128–214	1.39–2.33	Melack (1979)
Oloiden (Kenya)	98–281	146–420	1.58–4.54	Melack (1979)
Sibaya (South Africa)	5–26	–	0.23–1.85	Allanson (1979)
Swartvlei[a] (South Africa)	5–13	13–37	–	Robarts (1976)
Winam Gulf (Kenya)	86–240	150–341	1.61–3.68	Melack (1979)

[a] Estuarine-lake ecosystem.

concentration per unit volume was not a good indicator of areal productivity in the lake ($r = 0.41; n = 22$).

Changes in the photosynthetic capacity, A_{max}/B (B = biomass as chlorophyll), of the Lake McIlwaine phytoplankton ranged between 4 and 21 mg C (mg chlo a)$^{-1}$ h^{-1} (Table 2). These values are high when compared with other studies. While temperature had a significant effect on A_{max}/B ($r = 0.60; p = > 0.01 < 0.02; n = 19$), a highly significant inverse relationship between A_{max}/B and ΣB, biomass per unit area, was calculated ($r = 0.75; p = < 0.001; n = 19$). Thus, when the phytoplankton population was low the cells would have spent a greater percentage of time in the light and this could have been the reason for the inverse relationship (Robarts, 1979).

The data from Robarts' study indicated, as had been suggested earlier, that light was a major factor regulating primary productivity in Lake McIlwaine.

Table 2 Summary of the principal variables calculated for Lake McIlwaine phytoplankton productivity experiments (after Robarts, 1979)

Date	ΣB mg m^{-2}	I'_0 cal cm^{-2} h^{-1}	I_k cal cm^{-2} h^{-1}	ε_{min} ln units m^{-1}	A_{max}/B mg C (mg chlo a)$^{-1}$ h^{-1}	ΣA mg C m^{-2} h^{-1}	$\Sigma\Sigma A$ g C m^{-2} d^{-1}
18.06.75	84.3	14.7	2.47	1.47	4.92	279.0	2.40
02.07.75	76.3	23.3	2.86	1.42	7.47	375.7	2.99
23.07.75	94.3	24.8	3.47	1.43	6.49	333.6	2.69
07.08.75	118.5	26.3	1.92	1.43	4.03	304.9	2.69
27.08.75	93.6	25.4	3.02	1.50	5.45	333.6	2.87
03.09.75	—	—	—	—	7.47	451.3	—
24.09.75	86.9	31.1	2.52	1.95	6.38	425.9	3.93
08.10.75	—	—	—	—	7.29	586.9	—
28.10.75	101.3	16.5	1.21	2.91	6.19	429.1	4.05
05.11.75	120.8	34.4	3.61	2.86	6.83	517.3	4.86
24.11.75	76.6	30.6	2.73	1.45	8.53	457.1	4.79
18.12.75	72.2	17.6	1.42	2.48	11.67	567.1	5.94
14.01.76	32.9	38.9	2.30	1.45	21.29	599.6	6.03
28.01.76	48.0	32.2	3.86	1.04	13.54	411.2	3.63
11.02.76	108.3	—	—	1.52	10.27	561.5	—
25.02.76	134.3	34.2	3.25	2.30	6.82	635.1	5.54
10.03.76	101.2	26.6	4.28	2.55	7.58	489.1	4.15
23.03.76	95.7	31.4	3.14	2.16	9.12	652.7	5.60
07.04.76	72.5	18.9	3.85	1.46	6.70	248.1	1.64
21.04.76	115.6	21.6	4.77	1.25	8.19	461.3	3.95
05.05.76	126.6	20.8	2.37	1.46	4.89	389.6	3.09
19.05.76	101.4	30.1	4.51	1.16	6.44	416.4	3.43

While Robarts did not note a reduction in primary production due to the sewage diversion programme, more recent work by Thornton (1980) has indicated a significant improvement in water quality. He indicated that the trophic status of the lake had been reduced from hypereutrophic to eutrophic bordering on mesotrophic. Although his study did not include the measurement of primary production, recent measurements of productivity by Nduku (unpublished) suggest that even with the improvements in water quality noted by Thornton production has not significantly decreased. Robarts found that a combination of increased photosynthetic capacity and increased light penetration with smaller standing crops gave integral productivity values similar to values obtained with significantly greater algal population.

References

Allanson, B. R., 1979. The phytoplankton and primary productivity of the lake. In: B. R. Allanson, Lake Sibaya. Monogr. Biol., 36: 75–87.
Berman, T., 1976. Release of dissolved organic matter by photosynthesizing algae in Lake Kinneret, Israel. Freshwat. Biol., 6: 13–18.
Falconer, A. C., 1973. The phytoplankton ecology of Lake McIlwaine, Rhodesia. M. Phil. thesis, University of London.
Findenegg, I., 1964. Types of planktonic primary production in the lakes of the eastern Alps as found by the radioactive carbon method. Verh. Internat. Verein. Limnol., 15: 352–359.
Ganf, G. G., 1975. Photosynthetic production and irradiance – photosynthesis relationships of the phytoplankton from a shallow equatorial lake (Lake George, Uganda). Oecologia (Berl.), 18: 165–183.
Horne, A. J., 1979. Management of lakes containing N_2-fixing blue-green algae. Arch. Hydrobiol. Beih., 13: 133–144.
Lemoalle, J., 1973. L'energie lumineuse et l'activité photosynthétique du phytoplancton dans le Lac Tchad. Cah. O.R.S.T.O.M., sér. Hydrobiol., 7: 95–116.
Lewis, W. M. Jr., 1974. Primary production in the plankton community of a tropical lake. Ecol. Monogr., 44: 377–409.
Marshall, B. E. and A. C. Falconer, 1973a. Physico-chemical aspects of Lake McIlwaine, Rhodesia, a eutrophic tropical impoundment. Hydrobiol., 42: 45–62.
Marshall, B. E. and A. C. Falconer, 1973b. Eutrophication of a tropical African impoundment (Lake McIlwaine, Rhodesia). Hydrobiol., 43: 109–123.
Melack, J. M., 1976. Primary productivity and fish yields in tropical lakes. Trans. Amer. Fish. Soc., 105: 575–580.
Melack, J. M., 1979. Photosynthetic rates in four tropical African fresh waters. Freshwat. Biol., 9: 555–571.
Mitchell, D. S. and B. E. Marshall, 1974. Hydrobiological observations on three Rhodesian reservoirs. Freshwat. Biol., 4: 61–72.
Munro, J. L., 1966. A limnological survey of Lake McIlwaine, Rhodesia. Hydrobiol., 28: 281–308.
Munro, J. L., 1967. The food of a community of East African freshwater fishes. J. Zool., Lond., 151: 389–415.

Nduku, W. K., 1976. The distribution of phosphorus, nitrogen and organic carbon in the sediments of Lake McIlwaine, Rhodesia. Trans. Rhod. Scient. Ass., 57: 45–60.

Robarts, R. D., 1976. Primary productivity of the upper reaches of a South African estuary (Swartvlei). J. exp. mar. Biol. Ecol., 24: 93–102.

Robarts, R. D., 1979. Underwater light penetration, chlorophyll a and primary production in a tropical African lake (Lake McIlwaine, Rhodesia). Arch. Hydrobiol., 86: 423–444.

Robarts, R. D. and G. C. Southall, 1975. Algal bioassays of two tropical Rhodesian reservoirs. Acta hydrochim. hydrobiol., 3: 369–377.

Robarts, R. D. and G. C. Southall, 1977. Nutrient limitation of phytoplankton growth in seven tropical man-made lakes, with special reference to Lake McIlwaine, Rhodesia. Arch. Hydrobiol., 79: 1–35.

Robarts, R. D. and P. R. B. Ward, 1979. Vertical diffusion and nutrient transport in a tropical lake (Lake McIlwaine, Rhodesia). Hydrobiol., 59: 213–221.

Schmidt, G. W., 1973. Primary production of phytoplankton in three types of Amazonian waters. III. Primary productivity of phytoplankton in a tropical flood-plain lake of central Amazonia, Lago do Castanho, Amazonas, Brazil. Amazoniana, 4: 379–404.

Stewart, W. D. P., 1974. Blue green algae. In: A. Quispel, The biology of nitrogen fixation. Elsevier, New York.

Talling, J. F., 1965. The photosynthetic activity of phytoplankton in East African lakes. Int. Revue ges. Hydrobiol., 50: 1–32.

Thornton, J. A., 1979a. Some aspects of the distribution of reactive phosphorus in Lake McIlwaine, Rhodesia: phosphorus loading and seasonal responses. J. Limnol. Soc. Sth. Afr., 5: 33–38.

Thornton, J. A., 1979b. Some aspects of the distribution of reactive phosphorus in Lake McIlwaine, Rhodesia: phosphorus loading and abiotic responses. J. Limnol. Soc. Sth. Afr., 5: 65–72.

Thornton, J. A., 1980. Factors influencing the distribution of reactive phosphorus in Lake McIlwaine, Zimbabwe. D.Phil. Diss., University of Zimbabwe.

An examination of phytoplankton nutrient limitation in Lake McIlwaine and the Hunyani River system
Colleen J. Watts

Robarts and Southall (1977) looked at factors regulating phytoplankton growth in seven man-made lakes in Zimbabwe. Their study was carried out with a view to detecting the trophic status of the lakes and the nutrient limitation of phytoplankton growth in the lake waters. Because of its importance as Salisbury's principal water supply attention was focused on Lake McIlwaine, which received some sewage effluent and urban and industrial run-off, and it was compared with other lakes which do not receive sewage effluent. Robarts and Southall (1977) showed that Lake McIlwaine had nitrogen as the primary phytoplankton growth-limiting nutrient and the lake was thus designated as eutrophic. Other lakes principally showed phosphorus to be limiting and were classified as oligotrophic or mesotrophic.

The study of Watts (1980) was an extension of the work undertaken by Robarts and Southall (1977). In the past few years the Salisbury City Council has undertaken a sewage diversion programme which diverts treated sewage onto farmlands as fertiliser. Her study of the Hunyani River system (Fig. 8) including its tributaries and impoundments was undertaken in order to evaluate the reduction, if any, in the algal growth potentials (AGP's) of the water and any changes in growth-limiting nutrients that may have occurred as a result of the diversion of nutrients. It was also felt that seasonal variations in AGP's may be evident due to the rainfall pattern, and hence this aspect was studied in considerable detail.

Subsequent to the study of Robarts and Southall (1977) the Darwendale Dam impounding Lake Robertson was completed. Watts (1980) also looked at the AGP's and limiting nutrients of this new lake and compared it with Lake McIlwaine. AGP's and limiting nutrients were determined using the method of Robarts and Southall (1977); other methods are given in Watts (1980).

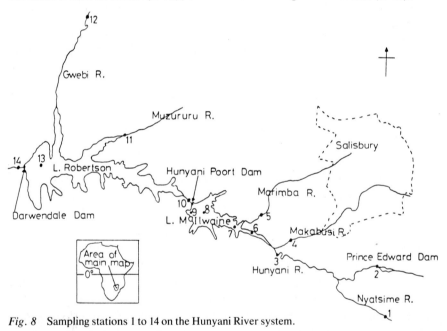

Fig. 8   Sampling stations 1 to 14 on the Hunyani River system.

The upper catchment

The upper catchment consists of those rivers flowing into Lake McIlwaine: namely, the Hunyani, Nyatsime, Makabusi and Marimba (Fig. 8). Both the Hunyani River (station 3; Fig. 9b) and the Nyatsime River (station 1; Fig. 9a) demonstrated considerable AGP's during the three seasons. The Makabusi

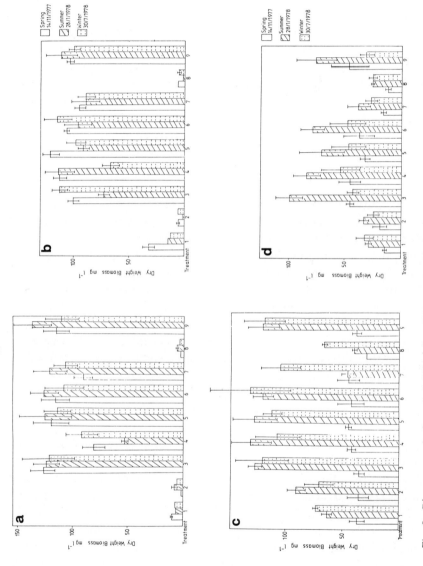

Fig. 9 Bioassay results from the [a] Nyatsime, [b] Hunyani, [c] Makabusi, and [d] Marimba Rivers. See appendix for key to treatments.

River (station 4; Fig. 9c) showed very little growth in the spring sample but considerable growth response in the winter and summer samples. The Marimba River (station 5; Fig. 9d), by comparison, shows much lower algal growth responses in all seasons, but this is most noticeable in the spring and winter.

The Nyatsime River, with the exception of the summer sample, and the Hunyani River have phosphorus as the primary growth-limiting nutrient and nitrogen as the secondary. It can be postulated therefore that agricultural run-off affects the nutrient levels of the water samples more than urban run-off. Both stations show some degree of tertiary sulphur limitation. This is due to the granite-derived soils in the catchment area (see R. S. Hatherly and K. A. Viewing, this volume) which are characterised by low sulphur levels. The more marked deficiency of sulphur in the summer may be caused by considerable dilution of sulphur by heavy rains.

In the Makabusi River, algal bioassays of the spring sample elucidated no primary limiting nutrient. The algal bioassays of the summer sample show a micro-nutrient as being primarily limiting with nitrogen as the secondary and phosphorus as the tertiary limiting nutrients. The fact that nitrogen is more limiting than phosphorus may be an indication that there has been some sewage spillage. In the winter sample there is no micro-nutrient deficiency and both nitrogen and phosphorus act as primary limiting nutrients. It is possible that the micro-nutrients were removed by filtration (Watts and Nduku, 1980). In the summer sample the micro-nutrient concentration may have been diluted by rainwater to have reached the threshold where the concentration was reduced by filtration sufficiently to induce deficiency. Rainfall dilution would be considerably lower during the winter (Table 3; note the increasing concentration of the ions) and hence micro-nutrient concentrations would not have been so critically affected by filtration.

Phosphorus, nitrogen and micro-nutrients are equally limiting in the Marimba River. It is again possible that filtration is responsible for the micro-nutrient deficiency. However, despite this, the possibility of true micro-nutrient deficiency should not be excluded out of hand. Robarts and Southall (1977) showed iron limitation on Lake Kariba in a water sample which had only been filtered through a $120 \mu m$ membrane filter which, according to the authors, does not reduce biologically available iron.

The reasons for the reduced AGP's at stations 4 and 5 are not clear. Chemical analyses of the water samples show nutrient levels to be high (Table 3). Theoretically algal growth should be very high therefore. Best growth was obtained with the summer sample for the Marimba River and with the summer and winter samples for the Makabusi River. Table 3 shows these waters

Table 3 Some physico-chemical characteristics of water samples from the upper catchment of Lake McIlwaine, including Prince Edward Dam (stations 1 to 5); concentrations in mg l^{-1} (except pH)

Parameter	Season	Station 1 Nyatsime	Station 2 P.E. Dam	Station 3 Hunyani	Station 4 Makabusi	Station 5 Marimba
pH	spring	7.1	7.05	7.25	8.15	7.75
	summer	6.35	6.5	6.5	6.5	6.7
	winter	7.1	6.8	6.9	7.2	7.75
Alkalinity	spring	49	27	45	115.5	158
($CaCO_3$)	summer	36	24	25	44	107
	winter	43.5	25	37	63.5	201
NO_3-N	spring	0.045	0.028	0.460	1.220	0.060
	summer	0.040	0.118	0.105	0.820	0.395
	winter	0.018	0.018	0.060	0.465	0.213
NH_4-N	spring	0.020	0.029	0.016	0.117	0.028
	summer	0.009	0.007	0.009	0.110	0.006
	winter	0.009	0.016	0.006	0.254	0.123
PO_4-P	spring	0.008	0.002	0.008	0.214	0.365
	summer	0.005	0.006	0.010	0.132	0.126
	winter	0.006	0.005	0.004	0.209	0.186
K	spring	3.82	3.20	4.30	12.45	7.00
	summer	1.68	2.05	2.55	7.40	4.25
	winter	1.33	1.50	1.65	5.77	5.34
Na	spring	10.95	8.70	12.80	54.25	42.25
	summer	5.82	4.48	4.80	17.64	16.30
	winter	8.60	5.50	7.00	40.00	40.00
Ca	spring	1.70	2.30	2.30	12.10	21.00
	summer	1.00	0.70	0.70	14.10	10.00
	winter	1.98	1.63	2.06	15.60	24.60
Mg	spring	2.00	0.92	1.70	6.15	10.45
	summer	1.14	0.68	0.72	3.65	5.42
	winter	2.26	1.26	1.86	10.00	13.70
Mn	spring	<0.02	0.02	0.04	<0.02	<0.02
	summer	<0.02	0.02	<0.02	<0.02	0.02
	winter	0.07	0.04	0.02	<0.02	0.02
Fe	spring	0.25	0.05	0.10	0.05	0.05
	summer	0.20	0.10	0.20	0.05	0.05
	winter	0.58	0.14	0.25	<0.01	0.02

to be, on average, lower in nutrients than the spring samples. The lower chemical concentrations would be due to dilution during the summer rains. It seems likely therefore that algal growth is being inhibited by some unidentified factor in the water. One possibility is that the high sulphate levels which are found periodically in both rivers may reduce the AGP's. Increasing sulphate concentrations have been shown to produce significant reductions in the AGP's (Watts, 1980). However, the reduction does not seem sufficient to account for the very marked drop in algal growth potentials obtained in the bioassays presented in Figs. 9c and 9d. The high levels of NaCl found at times in these rivers were also investigated by Watts (1980) as a possibility but there was no significant effect on the AGP caused by increasing the levels of NaCl in the culture medium. Another aspect which then must be considered is that the growth inhibition is species specific. Reduced algal growth is found with the culture alga, *Selenastrum capricornutum* Prinz, but examination of chlorophyll *a* concentrations show that at times there is considerable algal growth at both of these stations (Watts, 1980) which would indicate that there are algae which are capable of growing in these waters.

The lower catchment

The lower catchment consists of those rivers, flowing out of Lake McIlwaine and into and out of Lake Robertson. These rivers are the Hunyani, Muzururu and Gwebi Rivers (Fig. 8). Station 10 on the Hunyani River is situated just below the Hunyanipoort Dam spillway and could be expected to show similar results to Lake McIlwaine in terms of AGP's (see below). However, examination of Fig. 10a shows substantially different results. Although chemical analysis of the water sample shows it to be similar to the Lake McIlwaine samples (although NO_3-N and NH_4-N are for the most part actually higher), spring and summer samples produce much lower AGP's than anticipated. Only the winter sample produces results that compare in any way with Lake McIlwaine. An examination of chlorophyll *a* values (Watts, 1980) shows that water taken from this point is quite capable of maintaining a high phytoplankton standing crop since peaks of chlorophyll *a* are as high if not higher than those in water samples from the lake. It is possible that a heavy algal bloom might result in the water temporarily being depleted of nutrients, thereby biasing the results of the algal bioassays based on water sampled at that time, but this is unlikely since Lake McIlwaine does not show a lowering of the AGP during summer. Station 10 is also being continuously fed by water spilling over the dam wall at this time of year. Since station 10 is not affected by industrial and urban run-off nor by agricultural run-off to any extent, it is difficult to postulate any reason for the reduced AGP in the summer sample.

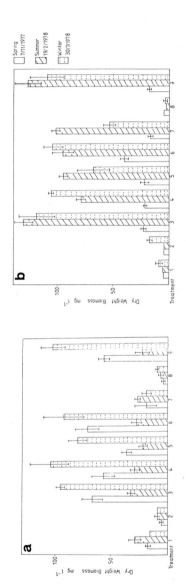
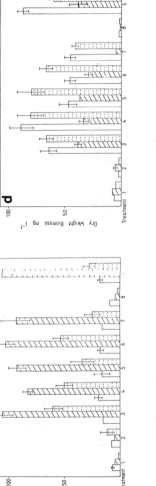

Fig. 10 Bioassay results from the [a] Hunyani, [b] Muzururu, [c] Gwebi, and [d] Hunyani Rivers. See appendix for key to treatments.

123

Table 4 Some physico-chemical characteristics of water samples from the lower catchment below Lake McIlwaine, including Lake Robertson (stations 10 to 14); concentrations in mg l^{-1} (except pH)

Parameter	Season	Station 10 Hunyani	Station 11 Muzururu	Station 12 Gwebi	Station 13 Lake Robertson	Station 14 Hunyani
pH	spring	7.9	7.75	7.75	7.45	7.15
	summer	7.3	6.85	7.1	7.05	7.0
	winter	7.0	7.5	7.65	7.0	7.0
Alkalinity ($CaCO_3$)	spring	88	112	147	60	58.5
	summer	147.5	49	75	68.5	70
	winter	46.5	106	126	55.5	54.5
NO_3-N	spring	0.063	0.028	0.040	0.035	0.040
	summer	0.595	0.048	0.032	0.018	0.018
	winter	0.035	0.018	0.028	0.038	0.050
NH_4-N	spring	0.115	0.030	0.014	0.013	0.013
	summer	0.007	0.009	0.006	0.006	0.007
	winter	0.092	0.012	0.023	0.013	0.028
PO_4-P	spring	0.003	0.002	0.001	0.001	0.001
	summer	0.004	0.006	0.004	<0.001	0.001
	winter	0.001	0.004	0.005	0.002	0.003
K	spring	4.30	1.75	2.70	3.92	3.98
	summer	0.85	1.40	1.30	4.25	4.68
	winter	2.32	1.04	0.85	1.87	1.80
Na	spring	16.25	23.50	12.00	10.28	9.80
	summer	21.40	7.80	7.15	9.15	9.15
	winter	7.50	12.70	8.90	7.50	6.70
Ca	spring	10.30	9.30	4.50	3.50	3.70
	summer	13.40	1.35	3.40	4.10	4.60
	winter	5.85	5.93	5.25	4.43	3.96
Mg	spring	4.28	5.15	10.80	4.08	10.13
	summer	4.92	1.95	3.65	4.32	4.42
	winter	3.36	8.90	10.60	5.90	6.05
Mn	spring	<0.02	0.04	0.02	<0.02	0.02
	summer	0.04	<0.02	0.02	0.04	0.04
	winter	0.02	<0.02	0.02	0.15	0.02
Fe	spring	0.05	0.05	0.05	0.05	0.05
	summer	0.05	0.10	0.10	0.05	0.05
	winter	<0.01	0.03	0.02	0.12	0.14

The only possibility that is reasonable to consider is that surface lake water from the spillway acts to dilute the river water to such an extent as to induce a nutrient deficiency, particularly in terms of potassium (Table 4). Water samples taken from the Hunyani River at station 14 below the Darwendale Dam spillway also show low AGP's (Fig. 10d).

As well as the Hunyani River, Lake Robertson is fed by the Muzururu and Gwebi Rivers. Both are only of real consequence during the summer and possibly for part of the winter as flow is reduced to almost nil by the end of the dry season in the spring. There was no flow at all in the Muzururu River when the spring sample was taken. Algal growth obtained in the bioassays of Muzururu and Gwebi River water is shown in Figs. 10b and 10c respectively. The spring samples and the winter sample from the Gwebi River only support very little algal growth; there is a dramatic increase in AGP at both stations during the summer. The poor algal growth in the two rivers during spring may be due to different causes. In the Muzururu spring sample, as high Na:K ratio (13:1) may be the cause of the reduced AGP. Watts (1980) has postulated that a high Na:K ratio is unfavourable for the growth of *Selenastrum capricornutum* on the basis of the bioassay results from the Muzururu and Hunyani (station 10) Rivers. In the case of the Muzururu this possibility is corroborated by the fact that the spring AGP is enhanced by the omission of $NaHCO_3$ from the culture medium thereby reducing the Na:K ratio. In addition, in summer when the Na concentration has been diluted by heavy rains, normal growth of *Selenastrum* is observed; this dilution effect is presumably still extant in winter as the river was still flowing at the time of sampling. The reason for the depressed AGP obtained in the spring sample from the Gwebi River is not clear, but may be related to the proximity of the river to the Great Dyke with its serpentine soils. It is unlikely that the river water is contaminated with heavy metals (nickel and chromium) in sufficient quantities to inhibit algal growth as Ferreira (1973) found only trace amounts of these ions in the river. Addition of EDTA also did not enhance algal growth (Watts, 1980) which suggests that heavy metals are not affecting algal growth in the Gwebi River. More likely is the low Ca:Mg ratio (0.4:1) caused by the presence of magnesium-rich montmorillonite clays of the serpentine soils. Dilution of this high Mg concentration during summer again leads to enhanced growth.

Samples from the Hunyani River (station 10) show phosphorus as being the primary limiting nutrient; nitrogen and one or more of the micro-nutrients are secondarily limiting in the spring and winter samples. Calcium also appears to be slightly limiting in the spring and winter samples, presumably reflecting the lower Ca levels found in Lake McIlwaine. Phosphorus is also primarily limiting in the Hunyani River at station 14 during spring and winter; phosphorus and micro-nutrients are limiting in the summer. The fact that Lake Robertson acts as a nutrient sink (Thornton, 1980) means that the nutrient status at station 14 is much lower in general than at station 10 (Table 4). The exception is the Ca and Mg concentrations which are higher as a result of run-off from the Great Dyke. As a consequence of this lower nutrient status, AGP's at station 14 are relatively low compared with station 10. In both the Gwebi and Muzururu Rivers, nitrogen and phosphorus are equally limiting during spring and summer while, during winter, nitrogen is the primary growth-limiting nutrient.

Lake McIlwaine

The overall picture for Lake McIlwaine is obtained from the algal bioassays done on water from four sampling points. Seasonal variation in AGP is most apparent at stations 6 and 7 (SM-1 and SM-3) (Figs. 11a and 11b). Both stations show nitrogen limitation in the summer which can be explained by the greater volume of water coming into the lake from the Makabusi and Marimba Rivers. The relatively high nutrient loadings near stations 6 and 7 would create a local effect. During spring and winter when the volume of water entering the lake from these two rivers is greatly reduced the primary growth-limiting nutrient is phosphorus, or phosphorus and nitrogen acting equally. By the time the water reaches stations 8 and 9 (SM-4 and SM-7) it is better mixed. AGP's are more or less similar throughout the year with phosphorus or phosphorus and nitrogen together being the principal growth-limiting nutrients. There is some indication that in winter and spring there are one or more micro-nutrients acting as tertiary growth-limiting nutrients, but this is considered to be marginal and may be induced by filtration (Watts and Nduku, 1980). It is possible that these results could have been obtained if the samples had been taken during or just following an algal bloom. Steijn *et al.* (1975a) showed that a micro-nutrient became primarily limiting in Reitvlei Dam, South Africa, but stated that this was probably due to an algal bloom immediately prior to sampling and that the situation was unlikely to be important when considering eutrophication problems. However, chlorophyll *a* values at these stations do not necessarily support this hypothesis (Watts, 1980); although in some conditions a high phytoplankton standing crop can be effectively masked by zooplankton grazing (Claesson and Ryding, 1977). Stations 7 and 9 also show the possibility of calcium and magnesium limitation following a significant drop in the concentrations of these nutrients during the summer of 1976–77 (Watts, 1980).

A comparison of the results obtained by Watts (1980) with those of Robarts

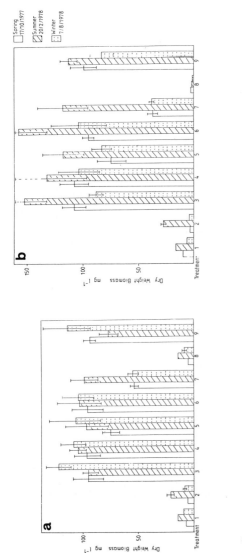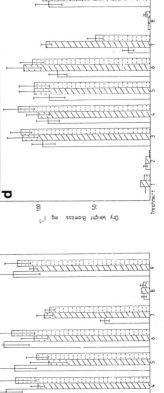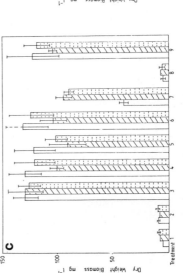

Fig. 11 Bioassay results from Lake McIlwaine stations [a] 6, [b] 7, [c] 8 and [d] 9. Station 8 is the mid-lake station. See appendix for key to treatments.

Table 5 Some physico-chemical characteristics of water samples from Lake McIlwaine (stations 6 to 9); concentrations in mg l^{-1} (except pH)

Parameter	Season	Station 6 (SM-1)	Station 7 (SM-3)	Station 8 (SM-4)	Station 9 (SM-7)
pH	spring	7.3	7.25	7.05	7.3
	summer	6.75	6.95	6.85	6.9
	winter	7.05	7.15	7.1	7.05
Alkalinity	spring	45	62	37.5	41
(CaCO$_3$)	summer	29	40	40	39
	winter	46.5	47	41.5	44
NO$_3$-N	spring	0.025	0.025	0.025	0.025
	summer	0.050	0.035	0.030	0.055
	winter	0.068	0.019	0.058	0.020
NH$_4$-N	spring	0.013	0.013	0.011	0.011
	summer	0.008	0.014	0.007	0.029
	winter	0.046	0.011	0.004	0.010
PO$_4$-P	spring	0.004	0.005	0.002	0.002
	summer	0.012	0.010	0.009	0.016
	winter	0.005	0.002	0.002	0.002
K	spring	3.75	3.10	3.10	4.00
	summer	2.70	2.95	3.20	3.10
	winter	2.38	2.18	2.15	2.32
Na	spring	11.90	9.80	9.65	11.70
	summer	5.55	8.25	8.28	8.26
	winter	9.80	10.20	7.60	9.00
Ca	spring	4.30	8.90	2.50	4.10
	summer	1.20	2.00	2.20	2.20
	winter	5.29	5.59	4.43	5.29
Mg	spring	2.10	1.90	1.80	2.10
	summer	0.95	1.65	1.72	1.65
	winter	3.18	3.30	3.36	3.17
Mn	spring	0.02	0.02	<0.02	0.02
	summer	0.02	0.02	0.02	0.04
	winter	<0.02	0.02	<0.02	<0.02
Fe	spring	0.05	0.05	0.05	0.05
	summer	0.10	0.10	0.05	0.05
	winter	0.02	0.01	0.04	0.01

and Southall (1977) shows that there has been an overall change at stations 8 and 9 (equivalent to stations 4 and Tiger Bay respectively of Robarts and

Southall) from nitrogen limited growth to phosphorus or phosphorus and nitrogen limited growth. Moreover, the actual biomass of algae obtained from the bioassays, conducted under similar conditions, has been reduced to approximately one third. This is corroborated by the work of Thornton (1979a) who showed that phosphorus loading to Lake McIlwaine has been reduced since the sewage diversion programme was initiated and that the lake acts as a phosphorus sink (Nduku, 1976; Thornton, 1979b; J. A. Thornton and W. K. Nduku, this volume).

Robarts and Southall (1975) state that the physico-chemistry of station 7 (their station 3) has always shown somewhat different characteristics to those of other Lake McIlwaine stations. This can be partially accounted for by the direction of the prevailing wind which causes the accumulation algae in this area. Certainly the algal bioassays results for station 7 during summer show markedly greater AGP's than at the other stations. Tiger Bay (station 9) also shows higher nutrient levels in some cases than the other Lake McIlwaine stations (Table 5). Beadle (1974) states that shallow waters close to shore lines are subject to different hydrobiological and productive regimes when compared with pelagic waters and that such areas are frequently more productive. The degree of seclusion of the bay and the direction of the mouth in relation to the winds and currents determine the extent to which production cycles in the bay and in the lake are interconnected. While Tiger Bay is secluded from winds blowing along the length of the lake, there is a large area of open water near the dam wall where the lake broadens and where winds blowing across the lake would affect Tiger Bay. Thus the anomalous AGP's observed at both stations 7 and 9 can potentially be ascribed to wind-induced circulation.

One of the unexplained features of the work of Robarts and Southall (1977) on Lake McIlwaine was the fact that with the algal bioassays from station 8 (equivalent to their station 4) reduced AGP's were obtained when all the nutrients were added (treatment 9; Robarts and Southall, 1977) compared with some of the other bioassay treatments. Likewise, Watts (1980) noted this anomaly in some of the bioassays undertaken in her study (Figs. 11a and 11b). Robarts and Southall (1977) suggested that this was caused by mixed glassware. While this is a possibility the repetition of the phenomenon during the study of Watts (1980) possibly negates this theory. It is possible that the reduced AGP's are caused by some form of chemical inhibition.

Prince Edward Dam and Lake Robertson
Prince Edward Dam (station 2; Fig. 12a) supports high algal growth only in the summer and winter; the spring sample supports a somewhat reduced

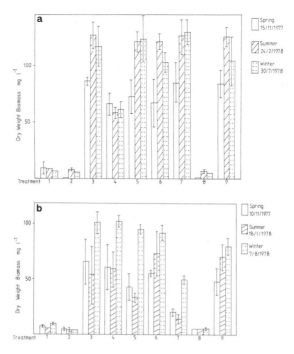

Fig. 12 Bioassay results from Prince Edward Dam [a] and Lake Robertson [b]. See appendix for key to treatments.

AGP. Phosphorus is indicated as the primary growth-limiting nutrient in spring. Both phosphorus and nitrogen equally enhance algal growth during the other seasons. This may indicate some urban run-off, especially during the rainy season, from the St. Mary's Township (see J. A. Thornton and W. K. Nduku, this volume). The algal bioassay results from the Prince Edward Dam summer sample compare closely to those obtained by Robarts and Southall (1977) during the same season. Robarts and Southall (1977) classified Prince Edward Dam as oligo-mesotrophic and the results of Watts (1980) would tend to corroborate this, on the basis of dual phosphorus and nitrogen limitation.

Figure 12b shows the seasonal algal growth responses obtained in the bioassays using water from Lake Robertson. During the spring and winter, inflows from the Gwebi River and Muzururu River have less effect on the lake than the Hunyani and the lake shows phosphorus as the primary growth-limiting nutrient. Station 13 is opposite the entrance of the Gwebi River and, during summer in particular, is influenced by the inflow from this river. This inflow alters the nutrient limitation to phosphorus and nitrogen limitation and there is some evidence of micro-nutrient and calcium deficiency. As in Lake

McIlwaine, there has been a major drop in calcium concentrations since 1976–77 (Watts, 1980) which is probably exacerbated by the low calcium concentration in the waters from the Gwebi River. Algal growth potentials are generally considerably lower than those obtained in Lake McIlwaine due to the removal of phosphorus in the upstream impoundment (Thornton, 1980). Lower chlorophyll a concentrations confirm this reduced growth potential in Lake Robertson (Watts, 1980).

It is interesting at this point to compare Figs. 12a and 10d which represent algal growth obtained from bioassays from Prince Edward Dam, the sampling station furthest upstream on the Hunyani River, and from station 14, the station furthest downstream. Overall AGP's have been reduced considerably, particularly during summer and winter. This conclusion is perhaps surprising in view of the large amount of nutrients added to the water in its passage through Lake McIlwaine, and would appear to justify the theory that the lakes act as sinks for nutrients.

Trophic status of the Hunyani River impoundments
The development of criteria to classify the trophic status of water bodies has been difficult. One method has been to examine one or two parameters only while another adopts a multi-parameter approach. Either approach has its drawbacks and in some cases waters may be classified as oligotrophic by one system and eutrophic by the other. Thus there appears to be no one universal criterion or even combination of criteria by which trophic status can be confidently assessed (Walmsley and Toerien, 1977).

Walmsley and Toerien (1977), in a study of three South African reservoirs, found they could classify the lakes on the basis of their oxygen profiles, and on the basis of the concentrations of iron, manganese and ammonia as eutrophic and as oligotrophic on the basis of the AGP's and chlorophyll a concentrations. Since the reservoirs demonstrated both oligotrophic and eutrophic characteristics they were tentatively classified as mesotrophic. On the other hand, Toerien *et al.* (1975) used the single parameter of AGP to assess trophic status in 98 major South African impoundments and found the AGP values to be a useful guide to trophic status in most cases. A number of studies in South Africa have used AGP's to classify lakes (Toerien and Steÿn, 1975; Steÿn *et al.*, 1975a, 1975b, 1976a, 1976b; Walmsley *et al.*, 1978). These studies have had equal success.

In Zimbabwe, studies of trophic status have been basically limited to one or two parameters, usually chemical characteristics or phytoplankton (population density and type or chlorophyll a). Only in work done by Robarts and Southall (1977) has any attempt been made to classify lakes on the basis of

AGP's. A survey of the trophic status of various Zimbabwean lakes has been described by Watts (1980). Due to the monomictic nature of most southern African lakes, the highest nutrient concentration at the surface may be expected during winter (Toerien *et al.*, 1975) and hence it is feasible that the best estimates of trophic status by the AGP method can be obtained during that season. Thus, using the winter lake water samples, the following trophic status classification can be made: Prince Edward Dam can be classified as mesotrophic, Lake McIlwaine as mesotrophic, and Lake Robertson as oligotrophic. Lake McIlwaine, previously classified as eutrophic (Robarts and Southall, 1977) is now classified as mesotrophic and shows an overall improvement in trophic status following nutrient diversion.

Nevertheless, with the increasing population of Salisbury and the consequent urban development and increased urban and industrial run-off, it may be difficult to maintain this improved trophic status and hence stringent measures may be required to ensure that this does not happen.

References

Beadle, L. C., 1974. Inland waters of tropical Africa. An introduction to tropical limnology. Longman, London.
Claesson, A. and S-O. Ryding, 1977. Nitrogen – a growth limiting nutrient in eutrophic waters. Prog. Wat. Tech., 8: 291–299.
Ferreira, J. C., 1973. A preliminary hydrobiological study of the rivers to be impounded by the Darwendale Dam. Hydrobiology Research Unit Rep., University of Rhodesia, Salisbury.
Nduku, W. K., 1976. The distribution of phosphorus, nitrogen and organic carbon in the sediments of Lake McIlwaine, Rhodesia. Trans. Rhod. Scient. Ass., 57: 45–60.
Robarts, R. D. and G. C. Southall, 1975. Algal bioassays of two tropical Rhodesian reservoirs. Acta hydrochim. hydrobiol., 3: 369–377.
Robarts, R. D. and G. C. Southall, 1977. Nutrient limitation of phytoplankton growth in seven tropical man-made lakes, with special reference to Lake McIlwaine, Rhodesia. Arch. Hydrobiol., 79: 1–35.
Steÿn, D. J., W. E. Scott, D. F. Toerien and J. H. Visser, 1975a. The eutrophication levels of some South African impoundments. I. Rietvlei Dam. Water SA, 1: 45–52.
Steÿn, D. J., D. F. Toerien and J. H. Visser, 1975b. Eutrophication levels of some South African impoundments. II. Hartbeespoort Dam. Water SA, 1: 93–101.
Steÿn, D. J., D. F. Toerien and J. H. Visser, 1976a. Eutrophication levels of some South African impoundments. III. Roodeplaat Dam. Water SA, 2: 1–6.
Steÿn, D. J., D. F. Toerien and J. H. Visser, 1976b. Eutrophication levels of some South African impoundments. IV. Vaal Dam. Water SA, 2: 53–57.
Thornton, J. A., 1979a. Some aspects of the distribution of reactive phosphorus in Lake McIlwaine, Rhodesia: phosphorus loading and seasonal responses. J. Limnol. Soc. Sth. Afr., 5: 33–38.
Thornton, J. A., 1979b. Some aspects of the distribution of reactive phosphorus in Lake McIlwaine, Rhodesia: phosphorus loading and abiotic responses. J. Limnol. Soc. Sth. Afr., 5: 65–72.

Thornton, J. A., 1980. A comparison of the summer phosphorus loadings to three Zimbabwean water-supply reservoirs of varying trophic states. Water SA, 6: 163–170.
Toerien, D. F. and D. J. Steÿn, 1975. The eutrophication levels of four South African impoundments. Verh. Internat. Verein. Limnol., 19: 1948–1956.
Toerien, D. F., K. L. Hyman and M. J. Bruwer, 1975. A preliminary trophic status classification of some South African impoundments. Water SA, 1: 15–23.
Walmsley, R. D. and D. F. Toerien, 1977. The summer condition of three eastern Transvaal reservoirs and some considerations regarding the assessment of trophic status. J. Limnol. Soc. Sth. Afr., 3: 37–41.
Walmsley, R. D., D. F. Toerien and D. J. Steÿn, 1978. Eutrophication of four Transvaal dams. Water SA, 4: 61–75.
Watts, C. J., 1980. Seasonal variation of nutrient limitation of phytoplankton growth in the Hunyani River system, with particular reference to Lake McIlwaine, Zimbabwe. M. Phil. thesis, University of Zimbabwe.
Watts, C. J. and W. K. Nduku, 1980. Loss of nutrients from water samples by filtration and its effect on algal bioassay procedures. J. Limnol. Soc. Sth. Afr., 6: 77–81.

Appendix

Algal growth responses were measures using cultures of *Selenastrum capricornutum* after nine days of incubation in membrane-filtered water samples with differing enrichments. Vertical lines in Figs. 9 to 12 represent two standard errors of the mean indicating differences in the growth response at 95% probability. Two standard errors of less than 1 mg l^{-1} are not shown. Nutrient enrichments were the same as those of Robarts and Southall (1977):

treatment 1 = no NaNO$_3$;
treatment 2 = no K$_2$HPO$_4$;
treatment 3 = no MgCl$_2$;
treatment 4 = no MgSO$_4$;
treatment 5 = no CaCl$_2$;
treatment 6 = no NaHCO$_3$;
treatment 7 = no micro-nutrients;
treatment 8 = no enrichment at all; and,
treatment 9 = enrichment with all nutrients.

Zooplankton and secondary production
J. A. Thornton and Helen J. Taussig

Species composition

Relatively little is known about the zooplankton of Lake McIlwaine. Munro (1966) compiled the first species list of the zooplankton of the lake. He found *Ceriodaphnia dubia* Richard to be dominant for most of the year, particularly during late winter and spring (July to December) when populations of

Table 6 Species composition of cladoceran and copepodan zooplankton in Lake McIlwaine, 1962–63 and 1975 (after Munro, 1966, and Magadza, 1977a, respectively); x = present

Species	1962–63	1975
Cladocera		
Diaphanosoma excisum	x	
Diaphanosoma permamatum		x
Daphnia laevis	x	x
Daphnia lumholtzi	x	
Ceriodaphnia dubia	x	
Ceriodaphnia cornuta		x
Moina dubia	x	
Bosmina longirostris	x	x
Copepoda		
Calanoida		
Thermodiaptomus syngenes		x
Tropodiaptomus orientalis	x	x
Cyclopoida		
Tropocyclops prosinus	x	x
Mesocyclops leukarti		x
Macrocylops albidus		x
Thermocyclops emini	x	
Thermocyclops neglectus		x[a]

[a] Given in Magadza and Mukwena (1979).

between 50 and 275 × 10^3 m^{-3} were observed (Fig. 5, Munro, 1966). More recently Magadza (1977a) has examined the zooplankton of the lake and has noted the absence of many of the species described by Munro (1966), notably *Ceriodaphnia dubia*. Numerous other differences were observed (Table 6) but these were attributed mainly to sampling frequency and methodology (Magadza, 1977a). Unfortunately, Magadza (1977a) gives no indication of the relative abundance of the zooplankton species which he recorded from the lake except to say that *Thermodiaptomus syngenes* Keifer was more numerous than *Tropodiaptomus orientalis* Brady; the former being the first record of *Thermodiaptomus syngenes* north of the Tropic of Capricorn. A complete list of the zooplankton species recorded from Lake McIlwaine is given in the addendum.

The zooplankton assemblage of Lake McIlwaine recorded by Munro (1966) is similar to that found elsewhere in southern Africa in mesotrophic impoundments (Walmsley and Butty, 1980; Thornton and Cotterill, 1978; Mills, 1977) whilst that recorded by Magadza (1977a) appeared to be similar to that of a eutrophic impoundment (Walmsley and Butty, 1980). It is possible therefore that the changes in the species composition noted by Magadza (1977a) are due to changes brought about by the enrichment of the lake as well as to the methodological factors mentioned above. Species composition data for the post-diversion period is lacking, but preliminary data given by Magadza and Mukwena (1979) indicate that *Tropodiaptomus orientalis* is an important component of the zooplankton of the lake during summer, and is replaced in winter by *Thermodiaptomus syngenes*. At the time of their study, *Thermocyclops neglectus* (Sars) was identified as the dominant zooplankton when the natural zooplankton populations from Lake McIlwaine were cultured in the laboratory (Magadza and Mukwena, 1979). Thus, the extent to which nutrient diversion has affected the zooplankton populations is not yet clear.

Secondary production

No estimates of secondary production have been made for Lake McIlwaine. Magadza and Mukwena (1979) have, however, calculated the development periods for the post-embryonic stages of *Thermocyclops neglectus* (Sars) over a range of temperatures and suggest that their data may indicate the acclimatisation of the growth rates to local conditions. They note that development times for the species are shorter at higher temperatures (Table 7). This adaptation to the higher water temperatures of tropical lakes has been observed elsewhere in Africa (Burgis, 1970; Magadza, 1977b, unpublished) and suggests that secondary production in these lakes could be extremely high.

Table 7 Development periods in days for post-embryonic stages of *Thermocyclops neglectus* (Sars) at selected temperatures (after Magadza and Mukwena, 1979)

Temperature (°C)	Nauplii to copepodids	Copepodids to adults	Nauplii to adults
20°	6.6	4.4	11.0
25°	2.0	1.1	3.1
30°	3.4	3.7	7.1

Addendum

A list of zooplankton recorded from Lake McIlwaine.

1. Coelenterata
   Limnocnida sp.

2. Rotifera
   Brachionus calyciflorus Pallas
   Brachionus caudatus Barrais & Daday
   Brachionus sp.
   Keratella tropica (Apstein)
   Trichocerca lucristata (Gasse)
   Sychaeta sp.
   Polyarthra sp.
   Filinia sp.
   Herarthra mira (Hudson)
   Conochilus sp.

3. Cladocera
   Diaphanosoma excisum Sars
   Diaphanosoma permamatum Brehm.
   Daphnia laevis Birge
   Daphnia lumholtzi Sars
   Ceriodaphnia dubia Richard
   Ceriodaphnia cornuta Sars
   Moina dubia deGuerne & Richard
   Bosmina longirostris (Müller)

4. Copepoda
   Calanoida
     Thermodiaptomus syngenes Keifer
     Tropodiaptomus orientalis Brady
   Cyclopoida
     Tropocyclops prosinus Sars
     Mesocyclops leukarti (Claus)
     Macrocyclops albidus (Jurine)
     Thermocyclops emini (Mrazek)
     Thermocyclops neglectus (Sars)

5. Diptera
   Chaoborus sp.

References

Burgis, M. J., 1970. The effect of temperature on the development time of eggs of *Thermocyclops* sp., a tropical cyclopoid from Lake George, Uganda. Limnol. Oceanogr., 15: 742–747.

Magadza, C. H. D., 1977a. A note on Entomostraca in samples from three dams in Rhodesia. Arnoldia Rhod., 8 (14): 1–4.

Magadza, C. H. D., 1977b. Determination of development period at various temperatures in a tropical cladoceran, *Moina dubia* deGuerne and Richard. Trans. Rhod. Scient. Ass., 58: 24–27.

Magadza, C. H. D. and P. Z. Mukwena, 1979. Determination of the post-embryonic development period in *Thermocyclops neglectus* (Sars) using cohort analysis in batch cultures. Trans. Rhod. Scient. Ass., 59: 41–45.

Mills, M. L., 1977. A preliminary report on the planktonic microcrustacea of the Mwenda Bay area, Lake Kariba. Trans. Rhod. Scient. Ass., 58: 28–37.

Munro, J. L., 1966. A limnological survey of Lake McIlwaine, Rhodesia. Hydrobiol., 28: 281–308.

Thornton, J. A. and N. G. Cotterill, 1978. Some hydrobiological observations on five tropical African montane impoundments. Trans. Rhod. Scient. Ass., 59: 22–29.

Walmsley, R. D. and M. Butty, 1980. Limnology of some selected South African impoundments. Water Research Commission, Pretoria.

Aquatic macrophytes and *Eichhornia crassipes*
M. J. F. Jarvis, D. S. Mitchell and J. A. Thornton

Species composition and distribution

Munro (1966) defined the shoreline habitats of Lake McIlwaine in terms of the presence or absence of aquatic macrophytes, and the major species of macrophyte where the plants were present. Relatively few areas of the lake shore were free of macrophytes (Fig. 13), most of these being granite outcrops or steep sand and gravel shores in the main lake basin. Munro (1966) identified the major species of aquatic plants in the impoundment as *Phragmites* sp., *Typha* sp., *Aponogeton* sp., and *Nymphea caerula*. In the more riverine upper reaches of the lake, he noted extensive beds of *Polygonum senegalense* which extended some 30 to 40 m out into the lake. These stands were often associated with *Lagarosiphon major*, forming dense mats, although the latter was also distributed through other areas of the lake.

The distribution of the macrophyte species was often related to their method of attachment (Sculthorpe, 1967). The emergent macrophytes (*Phragmites* sp. and *Typha* sp.) dominated the shoreline flora of the lake. *Phragmites* sp. was largely confined to the supra littoral zone or area of damp ground above the water line (Wetzel, 1975). *Typha* sp. covered the eulittoral and upper littoral zones of the shoreline, extending to a depth of about 3 m into the lake basin (Munro, 1966). *Polygonum* sp. filled this niche in the riverine portions of the lake. The middle and lower littoral zones supported stands of the floating-leaved macrophyte, *Aponogeton* sp., and, in the calm wind-sheltered areas, stands of *Nymphea* sp. These latter plants extended to some 4 to 5 m depth (Munro, 1966). Of the 74 km of shoreline, Munro (1966) estimated that the *Typha–Aponogeton* community extended over some 20% of the shoreline length, the *Nymphea* community over 40%, and the *Polygonum* community over 25% (Fig. 13). No mention of the freely floating macrophyte, *Eichhornia crassipes*, was made in Munro's study although unpublished reports of the Department of National Parks and Wild Life Management show that the plant was present in the lake at that time.

This distribution of aquatic macrophytes has changed considerably since 1963, possibly as a result of measures taken to control the spread of *Eichhornia crassipes* (see below). *Typha* sp. stands in the lake basin were greatly reduced following implementation of water hyacinth control measures in

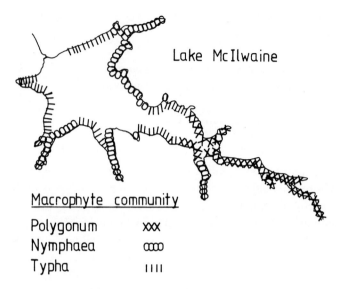

Fig. 13 The distribution of the macrophyte communities discussed in the text (after Munro, 1966).

1971 and *Nymphea* beds were decimated. *Phragmites* stands, on the other hand became slightly more abundant (M. J. F. Jarvis, personal observation) and *Polygonum* has also increased steadily since Munro's study. Ferriera (1974) recorded that this macrophyte can be found over about 90% of the lake shoreline (Fig. 14). She also recorded the presence of *Myriophyllum* sp. in the lake.

Munro (1966) and Ferreira (1974) agree that the *Polygonum* community is the most extensive of the macrophyte communities in Lake McIlwaine. Marshall (1971) noted that the rapid spread of this plant can be attributed in part to the break up of the large mats of *Polygonum* in the riverine portions of the lake which were then carried into the main lake basin (Fig. 14). These mats were often associated with *Eichhornia crassipes* (Ferreira, 1974).

Autecology of Polygonum senegalense

Ferreira (1974) investigated the autecology of the emergent macrophyte *Polygonum senegalense* on the basis of this plant's dominance of the macrophyte flora of the lake. She found that the distribution of the plant was related to water level and seed availability. Conditions required for germination of

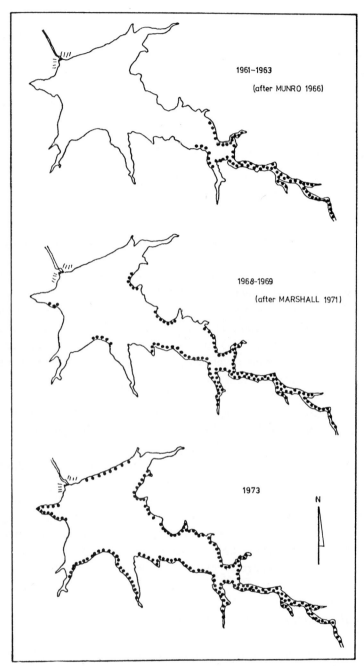

Fig. 14 The spread of *Polygonum senegalense* community at Lake McIlwaine since 1961 (after Ferreira, 1974).

Polygonum seedlings in Lake McIlwaine were found to be similar to those required by the temperate species of the plant, and were related to soil moisture and to a lesser degree to ambient temperature. She listed the ecological factors responsible for the success of *Polygonum senegalense* as follows:
1. the prolific production of seeds which remain dormant for prolonged periods and then germinate under unfavourable moisture conditions;
2. the rapid vegetative growth of the plant in a wide variety of habitats under favourable moisture conditions;
3. the tolerance of great fluctuations in water level;
4. the morphological adaptations to various aquatic habitats and the several life forms;
5. the effective distribution of seeds and vegetative reproduction;
6. the ability of the plant to die back and of the buds to remain dormant until moisture conditions become more favourable; and,
7. the tolerance of a wide range of soil characteristics such as texture, organic matter content, and conductivity.

She noted also that the gradual fluctuations of the seasonal cycle of the water level of the lake, resulting from the hydrological cycle in Zimbabwe (see B. R. Ballinger and J. A. Thornton, this volume), creates perhaps the ideal habitat for *Polygonum* by allowing for seedling germination on muddy substrates whilst ensuring that the mature plants remain in or near to the water.

The growth rate of the plant determined from field data obtained from Lake McIlwaine was on the order of 0.5 cm d^{-1} (Fig. 15). Germination was assumed to be on day 0 which was defined by the exposure of the mud substrate. The growth curve is only slightly sigmoidal with the lack of steepness of the slope of the growth phase of the curve possibly being a function of the soil moisture content. The plants flowered at a height of 60 cm under the conditions prevailing in Lake McIlwaine at the time of Ferreira's study. Under different environmental conditions, Ferreira (1974) has reported flowering at between 6 and 8 cm in height. These latter observations were made at Mazoe Dam where the substrate is very rocky and the soil moisture conditions very poor.

Eichhornia crassipes: management and control

Few of the aquatic macrophytes in Lake McIlwaine have caused any problems or have interfered with the use of the impoundment in any way. The exception to this generalisation is the water hyacinth, *Eichhornia crassipes*.

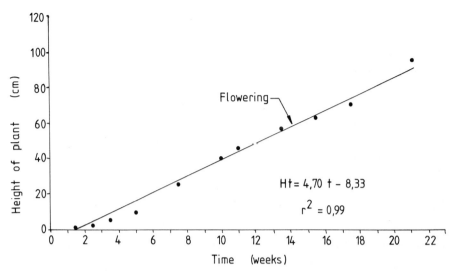

Fig. 15 Growth curve of *Polygonum senegalense* from Lake McIlwaine. Note the flat growth response which may be related to soil moisture content (after Ferreira, 1974).

Eichhornia crassipes is widespread throughout Zimbabwe (Gibbs-Russell, 1977; Jacot Guillarmod, 1979) and its presence in the Makabusi River system was recorded prior to the construction of Lake McIlwaine. The plant rapidly colonised the newly formed lake and was reported as 'widespread' on the lake surface by 1953 (Department of National Parks and Wild Life Management, unpublished). However, control was established by a combination of herbicide application and manual removal. Regular inspection patrols of the Makabusi River upstream of the lake were undertaken, and any plants that were found were removed and destroyed. For about a decade from the late 1950s, water hyacinth plants were rarely seen on the lake. However, following re-organisation of responsibilities within Government departments, the patrols were discontinued, and within about three years water hyacinth populations had increased to cover an estimated area in excess of 30% of the lake surface, with the largest concentrations being in Tiger Bay where the plants were swept by the wind. The increase in population was promoted by favourable conditions for seed germination following a period of low lake level, together with the plant's inherent capacity for rapid vegetative growth (in the spring of 1970 it was estimated that the population was doubling in size every 10 to 12 days).

A number of management options were considered to control the spread of *Eichhornia crassipes* on Lake McIlwaine (Mitchell, 1979a, 1979b). The three alternatives considered were mechanical control, chemical control and biological control. Biological control was not considered feasible due to the lack of a suitable indigenous parasite, although it was employed with some success against *Salvinia molesta* on Lake Kariba (Mitchell and Rose, 1979). Initially, mechanical control was employed to remove the plants from their shoreline habitats. These measures employed power-boats to push the weed into areas of shallow water where a net, pulled by a tractor, was used to haul the weed onto the shore whence it was removed for disposal. It was estimated that up to 100 tonnes of wet plant material was removed daily from the Tiger Bay area of Lake McIlwaine alone (Department of National Parks and Wild Life Management, unpublished). Macrophyte control by management of lake level was ruled out in the case of Lake McIlwaine as the impoundment was the primary water supply for the City of Salisbury. Obviously, other methods of control were required and the only one that seemed to offer a solution to the problem of water hyacinth infestation was chemical control.

The phenoxyl herbicide 2,4-D amine was selected as the control agent as this herbicide had been effective against large-scale aquatic weed infestations elsewhere. Data on the use of 2,4-D herbicide (Table 8) relies on the unpublished reports of the Hyacinth Control Officer (Department of National Parks and Wild Life Management, unpublished) and the volumes used were not always recorded. This herbicide is a phenoxyl based on butyl asopropyl amine 2,4-D at 47.76%. The emulsifying agent was 3% by weight and the hydrocarbon solvent 49.24%. The recommended application rate was 45 to 50 l ha^{-1} and it is effective against a variety of broadleaf plants.

At one time, this chemical was considered undesirable due to possible mutagenic properties (Tinker, 1971) and some wild life such as Lamellibranchia were found to contain high levels of residues that were possibly detrimental to the species. Trials were carried out to investigate the breakdown time of the active agent in the water and in the bottom muds. Aerial spraying of heavily infested sections of the lake was undertaken in stages as this procedure used much less chemical and proved to be more effective than manual spraying in many cases. In all spraying operations great care was taken and consideration give to the wind and current directions and other factors to ensure that there were no contamination problems in the irrigation and potable water supplies. Monitoring carried out during the spraying operation failed to detect any traces of 2,4-D amine in the raw water supplied for potable and irrigation usage, although the City of Salisbury placed activated carbon filters on line in the water works as an added precaution against contamination (Jarvis *et al.*, 1981).

The extensive use of 2,4-D helped to bring the water hyacinth problem

Table 8 The use of 2,4-D herbicide at Lake McIlwaine

Year	Litres used	Comments
1953	14 080	Water hyacinth widely spread
1954–55	14 080	Water hyacinth present
1955–56	14 080	
1956–57	14 080	
1957–58	7 500	Reed beds burnt to kill water hyacinth
1958–59	7 040	
1959–63	?	No records available for this period
1963–64	1 267	Water hyacinth still abundant
1964–65	nil	Water hyacinth greatly reduced
1965–66	619	
1966–67	1 408	Water hyacinth still widespread
1967–68	1 408	
1968–69	2 323	
1969–70	1 232	Water hyacinth abundant; aerial spraying undertaken
1970–71	3 801	Extensive aerial spraying twice around the lake
1971–72	1 480 minimum	Aerial spraying; little water hyacinth left
1972–73	?	No more floating water hyacinth left; manual spraying of shoreline patches
1973–74	?	Manual spraying of shoreline patches
1974–75	290 minimum	Manual spraying of shoreline patches
1975–76	290	Less used than last year
1976–77	?	Very little used
1977–78	nil	
1978–79	302	Patches on shore

under control during 1971 (Table 8). At the end of 1971 there were no more floating beds but only small amounts trapped in vegetation along the shore and a recurring problem with the germination of seedlings on exposed mud banks. This was, and is, controlled by manual spraying in shoreline areas.

When the water hyacinth disappeared it was noticed by many observers that nearly all of the *Nymphea* community and other aquatic vegetation had also gone. A reduction was also noted in the extent of the *Typha* community whilst the *Phragmites* reeds were more abundant. During the past three years (1978–81) there are signs that the *Nymphea* and *Typha* communities are beginning to expand again in some areas of the lake (M. J. F. Jarvis, personal observation).

Re-infestation of Lake McIlwaine is controlled by the provisions of the Parks and Wild Life (General) Regulations, 1975 (Rhodesia Government Notice 965 of 1975) which prohibits the spreading of aquatic weeds; namely,

Salvinia molesta (Kariba weed) and *Eichhornia crassipes* (water hyacinth).

The experience in the chemical control of the water hyacinth in Lake McIlwaine has been applied equally effectively elsewhere in southern Africa (Scott *et al.*, 1979).

References

Ferreira, J. C., 1974. Autecological studies of *Polygonum senegalense* Meisn. M.Sc. thesis, University of Rhodesia.
Gibbs-Russell, G. E., 1977. Keys to vascular aquatic plants in Rhodesia. Kirkia, 10: 411–502.
Jacot Guillarmod, A., 1979. Water weeds in southern Africa. Aquat. Bot., 6: 377–391.
Jarvis, M. J. F., M. I. van der Lingen and J. A. Thornton, 1981. Water hyacinth. Zimbabwe Sci. News, 15: 97–99.
Marshall, B. E., 1971. Ecology of the bottom fauna of Lake McIlwaine (Rhodesia). M.Sc. thesis, University of London.
Mitchell, D. S., 1979a. Assessment of aquatic weed problems. J. Aquat. Plant Manage., 17: 19–21.
Mitchell, D. S., 1979b. Formulating aquatic weed management programs. J. Aquat. Plant Manage., 17: 22–24.
Mitchell, D. S. and D. J. W. Rose, 1979. Factors affecting fluctuations in extent of *Salvinia molesta* on Lake Kariba. Pest. Artic. and News Summ. (PANS), 25: 171–177.
Munro, J. L., 1966. A limnological survey of Lake McIlwaine, Rhodesia. Hydrobiol., 28: 281–308.
Scott, W. E., P. J. Ashton and D. J. Steÿn, 1979. Chemical control of the water hyacinth on Hartbeespoort Dam. Water Research Commission, Pretoria.
Sculthorpe, C. D., 1967. The biology of aquatic vascular plants. Arnold, London.
Tinker, J., 1971. Unhealthy herbicides. New Scientist, 49: 593.
Wetzel, R. G., 1975. Limnology. Saunders, Philadelphia.

The benthic fauna of Lake McIlwaine
B. E. Marshall

The community of animals living on the lake bottom, the benthic fauna, is a complex one with representatives of almost every class of aquatic animal, and a greater variety of kinds and numbers than any other community (Reid, 1961). They play an important role as decomposers in the nutrient cycle of a lake, being largely detritivores and consumers of particulate organic matter. They are also important as fish food and can be useful indicators of pollution or eutrophication (Beeton, 1969; Johasson, 1969). In Lake McIlwaine the benthic fauna was investigated as part of a fishery survey (Munro, 1964, 1966) and in an attempt to evaluate the effects of eutrophication (Marshall, 1971,

1978). A severe drop in lake level in 1972–73 made it possible to assess the structure and abundance of the mussel population (Marshall, 1975).

Depth distribution and zonation

The principal sub-zones of a lake bottom are the littoral, or zone where rooted hydrophytes occur, the sub-littoral, and the profundal which is the area of the bottom beneath the thermocline. These zones can be distinguished in Lake McIlwaine by the depth distribution of the benthic fauna.

Two transects were made from the water's edge to the old river bed to show the effects of oxygen availability and water level fluctuations. The first was made in December 1968 when the water level had dropped 3.9 m and the lake was strongly stratified with no oxygen below 10 m. Both factors greatly influenced the distribution of the benthic fauna (Fig. 16a). No animals were found below 8 m of water because of the insufficient oxygen. This marks the beginning of the profundal zone where the water is deoxygenated throughout the summer months (Marshall and Falconer, 1973) and the point where the oxygen content of the mud reaches analytical zero (Nduku, 1976).

From the water's edge to the thermocline the benthic fauna was dominated by oligochaetes of which *Branchiura sowerbyi* Beddard was the most abundant species. Other oligochaetes included *Limnodrilus hoffmeisteri* Claparede and a small population (75 m^{-2}) of *Dero digitata* (Muller) which occurred at 8 m depth. The numbers of Chironomidae larvae were low and consisted almost entirely of *Procladius* sp. and other tanypods. The zone from the water's edge to the thermocline could be considered to be the sub-littoral. In Lake McIlwaine the littoral is probably better defined as the zone affected by water level fluctuations, which are usually about 2 m yr^{-1} (see B. R. Ballinger and J. A. Thornton, this volume). In December 1968 no littoral zone existed because of the exceptional drop in water level.

A second transect was made in August 1969 and illustrated the three zones more clearly (Fig. 16b). At this time the lake was still isothermal and oxygen occurred at all depths (although the concentration in the profundal did not exceed 40% saturation) and populations of *Dero digitata* and the chironomids *Procladius* sp. and *Chironomus* sp. were present. However their numbers were never greater than 1000 m^{-2} and the deepest muds, which were described as 'slimy and evil-smelling' (Munro, 1966), were generally unsuitable for benthic animals.

The sub-littoral extended from 4 m depth (previously the water's edge) to 12 m depth (previously 8 m) and was dominated by the oligochaete *Branchiura*

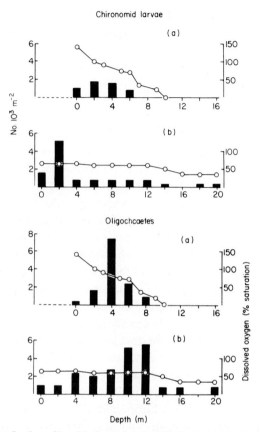

Fig. 16 Depth distribution of benthic fauna in Lake McIlwaine during (a) December 1968 and (b) August 1969. Dissolved oxygen concentrations (o) are also shown. Due to the low water level during December 1968, the depth scale has been shifted to the right; e.g., the site at the water's edge in December was 4 m deep in August (after Marshall, 1978).

sowerbyi. Chironomid numbers were low in this zone, never exceeding 1000 m^{-2} with Tanypodinae being the most important.

The littoral zone was now present and consisted of the previously exposed lake bed and extended from the water's edge to 4 m depth. An extensive growth of marginal vegetation (principally *Polygonum senegalense* Meisn. with *Typha latifolia* L. and *Phragmites mauritianus* Kunth. on the inner shoreline) had developed and an abundant chironomid fauna had become established. The most numerous of these were *Chironomus* sp and *Nilodorum* sp. which are able to colonise newly flooded land (McLachlan, 1969a, 1970, 1975).

The depth distribution of the benthic fauna in Lake McIlwaine is similar to that in other African impoundments and is apparently limited by the availability of dissolved oxygen. In Lake Kariba there were no chironomids below the thermocline, but a *Polypedilum* sp. occurred in the profundal after overturn when the bottom water had become oxygenated (McLachlan, 1969b). In the Volta Lake benthic animals did not occur under water with less than 30% oxygen saturation (Petr, 1969). It appears that 20% saturation in the water will limit benthic fauna in Lake McIlwaine.

The influence of lake level fluctuations

Changes in the benthic fauna of the littoral zone were investigated as this is the area where chironomid larvae are most abundant and these are an important fish food (Munro, 1967; Marshall and van der Heiden, 1977). Water level changes were the dominant factor in the littoral zone and observations were made during two drought periods (1967–68 and 1972–73) when the lake was unusually low. They were followed by heavy rains and a rapid rise in the water level.

The effects of this on the benthic fauna in 1968 are shown in Fig. 17. During the low water period the littoral was exposed; there was no marginal vegetation and the benthos consisted almost entirely of oligochaetes. *Branchiura sowerbyi* numbers reached 2000 m^{-2} whilst *Limnodrilus hoffmeisteri* was less abundant with numbers below 500 m^{-2}. Few chironomids were found with *Procladius* sp. being the most abundant.

This changed when the lake rose in December-January and flooded the exposed shore and new vegetation growing there. Oligochaete numbers dropped rapidly until they reached their lowest level in April (300 m^{-2}). The chironomids responded almost immediately and two genera were able to colonise the newly flooded land. *Chironomus* sp. and *Nilodorum* sp. had been nearly absent during the low water period but now became abundant and by February their total density was nearly 2000 m^{-2}. Their numbers remained high until July whilst the water level was stable, but declined rapidly as the water level dropped. *Cryptochironomus* sp. was the most abundant of the other chironomid genera.

The tanypod larvae showed a less distinct response to water level changes. *Procladius* sp. was the commonest of these and they were most numerous during August to October 1968 when the lake level was lowest. They were common again during July to October 1969 when the lake level was high but dropping. This suggests that they may be seasonally abundant and less dependent on the lake level.

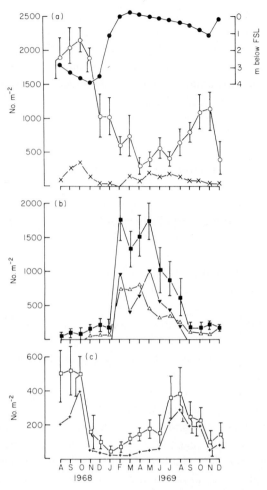

Fig. 17 The abundance of the main benthic fauna groups in the littoral zone; (a) oligochaetes (● = lake level, o = *Branchiura sowerbyi*, x = *Limnodrilus hoffmeisteri*, FSL = full supply level); (b) chironomids (▼ = *Chironomus* sp., ■ = total chironomids, △ = *Nilodorum* sp.); and, (c) Tanypodinae (□ = total Tanypodinae, + = *Procladius* sp.). Vertical lines represent 95% confidence limits of the mean but have been omitted from the less abundant groups for clarity (after Marshall, 1978).

Neither the chironomids nor the tanypods show a clear relationship between their numbers and the lake level (Fig. 18). In the Chironomidae, numbers were generally low when the lake level was rising and falling but were abundant whilst the water level was high and stable. The tanypods showed no relationship with lake level.

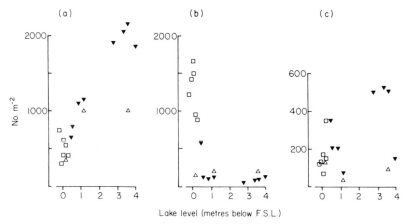

Fig. 18 The relationship between lake level and (a) oligochaetes, (b) chironomids and (c) tanypods when the lake is rising (△), falling (▼) and stable (□ = ± 0.2 m below FSL). Only the oligochaetes show a positive correlation ($r = 0.88$; $p < 0.01$) (after Marshall, 1978).

By contrast the oligochaetes showed a clear relationship between their abundance and lake level ($r = 0.879$). These animals were most abundant in the sub-littoral zone (Fig. 16) and were not able to move into the littoral as quickly as the chironomids. Thus a drop in lake level reduces the littoral zone and the numbers of oligochaetes will increase.

A comparison between 1962–63 (Munro, 1966) and 1968–69 (Marshall, 1978) suggests that chironomid numbers usually increase after a rise in lake level (Fig. 19). The lake was relatively stable during the early study but a slight rise in water level in December 1962 was followed by a dramatic increase in chironomids. They dropped rapidly soon afterwards although the lake level remained high. There was a slight increase in June and July 1963. In 1968–69 chironomid numbers increased in February after the lake level rose and remained high for a longer period.

Few other groups were recorded. A large population of lamellibranchs, mainly *Mutela dubia* (Gmelin), was stranded on the exposed lake bottom and none were taken during the sampling programme. Several other groups were found in low numbers after the lake level rose. The most abundant of these were bulinid snails (maximum population 53 m^{-2}) and a ceratopogonid larva (maximum population 57 m^{-2}). Other groups found included odonate and ephemeropteran nymphs, larvae of *Chaoborus edulis* Edwards and leeches.

The lamellibranchs were the group most affected by the severe drops in water level which took place in 1968–69 and 1972–73. These decimated the mussel population. Their numbers and distribution were recorded following the second drought when the lake dropped 5.1 m exposing virtually the entire population (Marshall, 1975).

The most common species was *Mutela dubia* which occurred at all stations except two by the dam wall where the shore was steep and rocky (Fig. 20a). The greatest density recorded was 31 m^{-2} in the Hunyani River section where the population was generally the highest. The mean throughout the lake was about 6 m^{-2} although 24 m^{-2} were noted on an exposed sandy shore at Pelican Point.

Caelature mossambicensis (Martens) was less abundant although also relatively common in the Hunyani River section where the greatest density recorded was 2.6 m^{-2} (Fig. 20b). Few *Corbicula africana* Krauss were found and then only in the Hunyani River section (Fig. 20c). This species occurs in greatest numbers amongst the marginal vegetation and was probably stranded early in the season.

The increase in chironomids following a water level rise was probably brought about by the nutrients released following submersion of plant growth. A similar phenomenon was noted in Lake Kariba after the inundation of vegetation and animal dung (S. M. McLachlan, 1970). In Lake Kariba the

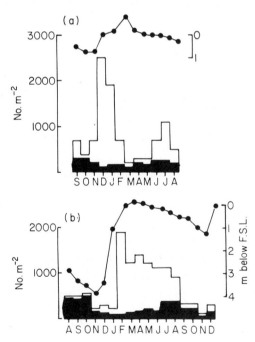

Fig. 19 The relationship between lake level and chironomids (open histogram) and tanypods (solid histogram) during 1962–63 (a) and 1968–69 (b) (after Munro, 1966; Marshall, 1978).

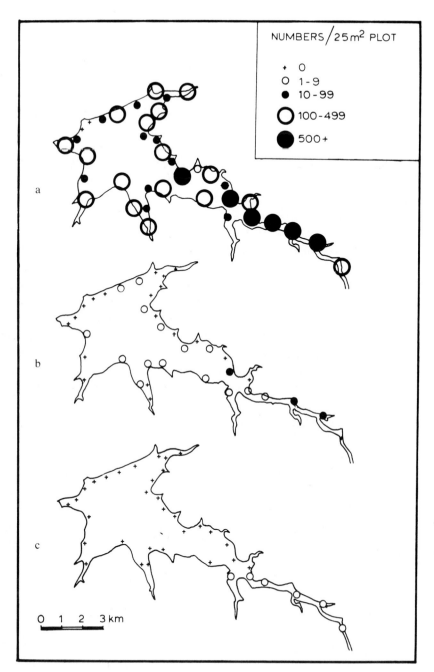

Fig. 20 The distribution and density of the three commonest mussels in Lake McIlwaine; [a] *Mutela dubia,* [b] *Caelatura mossambiciensis,* and [c] *Corbicula africana* (after Marshall, 1975).

horizontal movement associated with rising water considerable and large numbers of *Chironomus transvaalensis* Keiffer were found only in very shallow water (McLachlan, 1969b, 1970). By contrast, in Lake McIlwaine the water rose rapidly with little horizontal movement, and *Chironomus* and *Nilodorum* were found throughout the littoral zone. The nutrient-rich water and large quantities of flooded vegetation probably accounted for this. A similar increase in chironomids was noted in 1962–63 (Munro, 1966) but lasted for a shorter period possibly because of the less extreme water level changes and correspondingly smaller release of nutrients (Fig. 19).

In Lake McIlwaine the main colonisers were both *Chironomus* and *Nilodorum* which is in contrast to Lakes Kariba and Chilwa where *Chironomus transvaalensis* was almost exclusively the colonising species (McLachlan, 1969b, 1970). The reasons for this are not clear, but McLachlan (1975) has shown that *Chironomus transvaalensis* has a more rapid life cycle than *Nilodorum brevipalpis*, which may enable it to be the first to colonise newly flooded ground in gently sloping areas. This advantage was probably lost in Lake McIlwaine where the rise in water level was rapid and the exposed ground was quickly submerged. McLachlan (1975) also demonstrated that *Nilodorum brevipalpis* showed a preference for a mud substrate whilst *Chironomus transvaalensis* preferred plants. This suggests that the organically-rich sediments of Lake McIlwaine were more suitable for *Nilodorum*, which was thus able to compete successfully with *Chironomus* as a colonising species.

It has been suggested that annual fluctuations in lake level might have a permanent effect by reducing the diversity of animals living in the mud. This has been noted in man-made lakes in West Africa (Thomas, 1966) and Canada (Nursall, 1969) and is supported by the fact that African man-made lakes have a relatively poor fauna in the shallow areas influenced by such fluctuations whereas natural lakes have a much more varied one (McLachlan, 1974). This is the case in Lake McIlwaine where the benthic fauna consists almost solely of chironomids and oligochaetes, with lamellibranchs only becoming established during periods of stable water level.

Influence of macrophytes

One of the most striking features of Lake McIlwaine is the disappearance of the large *Nymphea* sp. beds which in 1962–63 extended along 42% of the shoreline (Munro, 1966). This has been attributed to the phytophagous fish, *Tilapia rendalli* Boulenger (formerly *Tilapia melanopleura*) (Junor, 1969) but may also be an effect of increasingly dense algal blooms that occurred during this period and the chemical control of *Eichhornia crassipes* (see M. J. F. Jarvis *et al.*, this volume). Submerged macrophytes have been shown to play an important role in the development of benthic populations (McLachlan, 1969b, 1975) and their disappearance may have had an effect in Lake McIlwaine.

The benthic fauna collected by Munro (1966) from different vegetation types in 1962–63 can be compared with that collected in 1968–69 (Table 9).

Table 9 Mean population of major benthic groups in relation to marginal vegetation types (No. m^{-2}) during 1962–63 (after Munro, 1966) and 1968–69 (after Marshall, 1971); + = present, − = absent

Benthic group	1962–63			1968–69
	Nymphea	Typha	Polygonum	All stations
Branchiura sowerbyi	2480	1350	1853	769
Limnodrilus hoffmeisteri[a]	+	+	+	141
Hirudinea	−	−	1	2
Ephemeroptera nymphs	−	1	−	1
Odonata nymphs	8	4	−	1
Tricoptera larvae	31	8	39	+
Chaoborus sp. larvae & pupae	11	2	88	+
Chironomidae larvae & pupae	1072	638	163	586
Tanypodinae larvae & pupae	222	263	136	156
Culicinae larvae & pupae	3	−	10	10
Corbicula africana[b]	200	228	536	−
Totals	4053	2529	2832	1687

[a] Probably the species referred to as 'unidentified Oligochaeta' by Munro (1966).
[b] Munro (1966) refers to *Sphaerium* sp. which appears to be a misidentification of *Corbicula africana*.

The most significant difference is the reduction in numbers of *Corbicula africana* which appears to be most abundant amongst vegetation in shallow water in the lake. Its loss could, however, be attributed to the drop in lake level in late 1968 rather than a lack of submerged macrophytes. The only other significant reductions were amongst the Odonata and Trichoptera which are both commonly associated with macrophytes.

The loss of these plants does not, therefore, appear to have significantly reduced benthic productivity. Water level changes would appear to be more significant especially on groups such as the lamellibranchs which can be almost totally eliminated by an exceptional drop in the lake level.

Effects of eutrophication

The effects of eutrophication on the benthos of the lake is less marked than the influence of water level changes. In the littoral zone *Chironomus* was abundant for a longer period in 1968–69 than in 1962–63 which may have been a result of greater nutrient availability. The appearance of *Limnodrilus hoffmeisteri*, which is common in organically polluted situations (Hynes, 1960; Brinkhurst, 1966; Johnson and Matheson, 1968; Johasson, 1969), may also be a consequence of eutrophication.

The presence of *Chironomus* in the profundal in winter also confirms the eutrophic state of the lake using Brundin's (1957) system of lake classification. This is based on the presence of chironomid larvae in the profundal and shows a progression from oligotrophic *Tanytarsus*-Orthocladiinae lakes to eutrophic *Chironomus* lakes. By contrast Lake Kariba was classed as a *Polypedilum* lake (McLachlan, 1969c). Work in the Makabusi River showed that *Chironomus* was the most abundant chironomid in an organically polluted situation (Marshall, 1972) and the high level of organic matter in the muds of Lake McIlwaine (Nduku, 1976) enables it to be the dominant species there.

Acknowledgement

This paper is published with the approval of the Director of National Parks and Wild Life Management, Zimbabwe.

References

Beeton, A. M., 1969. Changes in the environment and biota of the Great Lakes. In: National Academy of Sciences, Eutrophication: causes, consequences and correctives. National Academy of Sciences, Washington.
Brinkhurst, R. O., 1966. The Tubificidae (Oligochaeta) of polluted waters. Verh. Int. Verein. Limnol., 16: 854–859.
Brundin, L., 1957. The bottom faunistical lake type system and its application to the S. hemisphere. Verh. Int. Verein. Limnol., 13: 288–297.
Hynes, H. B. N., 1960. The biology of polluted waters. University of Liverpool Press, Liverpool.
Johnson, M. G. and D. H. Matheson, 1968. Macroinvertebrate communities of the sediments of Hamilton Bay and adjacent Lake Ontario. Limnol. Oceanogr., 13: 99–111.
Johasson, P. M., 1969. Bottom fauna and eutrophication. In: National Academy of Sciences, Eutrophication: causes, consequences and correctives. National Academy of Sciences, Washington.

Junor, F. J. R., 1969. *Tilapia melanopleura* Dum. in artificial lakes and dams in Rhodesia with special reference to its undesirable effects. Rhod. J. Agric. Res., 7: 61–68.

McLachlan, A. J., 1969a. Some effects of water level fluctuation on the benthic fauna of two Central African lakes. Limnol. Soc. Sth. Afr. Newslett., 13: 58–63.

McLachlan, A. J., 1969b. The effect of aquatic macrophytes on the variety and abundance of benthic fauna in a newly-created lake in the tropics (Lake Kariba). Arch. Hydrobiol., 66: 212–231.

McLachlan, A. J., 1969c. A study of the bottom fauna of Lake Kariba. Nuffield Kariba Res. Stn. Rep. 1962–68. University of Rhodesia, Salisbury.

McLachlan, A. J., 1970. Submerged trees as a substrate for benthic fauna in the recently created Lake Kariba (Central Africa). J. appl. Ecol., 7: 253–266.

McLachlan, A. J., 1974. Development of some lake ecosystems in tropical Africa with special reference to invertebrates. Biol. Rev., 49: 365–397.

McLachlan, A. J., 1975. The role of aquatic macrophytes in the recovery of the benthic fauna in a tropical lake after a dry phase. Limnol. Oceanogr., 20: 54–63.

McLachlan, S. M., 1970. The influence of lake level fluctuation on water chemistry in two gradually shelving areas in Lake Kariba, Central Africa. Arch. Hydrobiol., 66: 499–510.

Marshall, B. E., 1971. Ecology of the bottom fauna of Lake McIlwaine (Rhodesia). M.Phil. Thesis, University of London.

Marshall, B. E., 1972. Some effects of organic pollution on benthic fauna. Rhod. Sci. News, 6: 142–145.

Marshall, B. E., 1975. Observation on the freshwater mussels (Lamellibranchia: Unionacaea) of Lake McIlwaine, Rhodesia. Arnoldia Rhod., 7 (16): 1–16.

Marshall, B. E., 1978. Aspects of the ecology of benthic fauna in Lake McIlwaine, Rhodesia. Freshwat. Biol., 8: 241–249.

Marshall, B. E. and A. C. Falconer, 1973. Physico-chemical aspects of Lake McIlwaine (Rhodesia), a eutrophic tropical impoundment. Hydrobiol., 42: 45–62.

Marshall, B. E. and J. T. van der Heiden, 1977. The biology of *Alestes imberi* Peters (Pisces: Characidae) in Lake McIlwaine, Rhodesia. Zool. Afr., 12: 329–346.

Munro, J. L., 1964. Feeding relationships and production of fish in a Southern Rhodesian lake. Ph.D. Diss., University of London.

Munro, J. L., 1966. A limnological survey of Lake McIlwaine, Rhodesia. Hydrobiol., 28: 281–308.

Munro, J. L., 1967. The food community of East African freshwater fishes. J. Zool. Lond., 151: 389–415.

Nduku, W. K., 1976. The distribution of phosphorus, nitrogen and organic carbon in the sediments of Lake McIlwaine, Rhodesia. Trans. Rhod. Scient. Ass., 57: 45–60.

Nursall, J. R., 1969. Faunal changes in the oligotrophic Kananaskis River system. In: L. E. Obeng, Man-made lakes: the Accra symposium. University of Ghana Press, Accra.

Petr, T., 1969. Development of bottom fauna in the man-made Volta Lake in Ghana. Verh. Int. Verein. Limnol., 17: 273–281.

Reid, G. K., 1961. Ecology of inland waters and estuaries. Rheinhold, New York.

Thomas, J. D., 1966. Some preliminary observations on the fauna and flora of a small man-made lake in the West African savanna. Bull. Inst. Fond. Afr. Noir, 28: 542–559.

The fish of Lake McIlwaine
B. E. Marshall

Because of its proximity to Salisbury, Lake McIlwaine is an important recreational centre (see G. F. T. Child and J. A. Thornton, this volume) and the surrounding land was proclaimed as a National Park soon after the lake was formed. Angling is a major attraction and a commercial fishery was established in 1956. Conflict between the two forms of fish utilisation has frequently arisen, but the commercial fishery was justified on the grounds that undesirable angling species such as *Barbus, Labeo,* and *Clarias* would be removed (Stewart, 1957). More recent information suggests that the conflict between angling and commercial fishing is not severe (Marshall, 1978a).

A lucrative illegal fishery also operates but the quantity of fish taken is unknown. In addition, a large number of poorer people fish on a subsistence basis to meet their own needs. Fish in the lake have thus evoked much interest and research projects have been carried out by the University of Zimbabwe and the Department of National Parks and Wild Life Management. The latter established a Research Centre at Miller's Creek on the north bank of the lake in 1973.

A brief account of the first year of the commercial fishery was given by Lewin (1957) and Marshall (1978a) used data from the commercial fishery and from angling surveys to make an estimate of the fish yield from the lake. An earlier study of the fish populations emphasised food and feeding relationships (Munro, 1964, 1967) whilst the abundance and growth of juveniles in the marginal areas was assessed by Marshall and Lockett (1976).

Other studies have concentrated on single species in the lake. A review of *Tilapia rendalli* in Zimbabwe included a discussion of its role in the destruction of marginal vegetation (Junor, 1969) whilst its food, growth and metabolism was investigated by Caulton (1975, 1976). The biology of the important cichlid, *Sarotherodon macrochir**, is less well known except aspects of its breeding biology (Marshall, 1978b) and the ecology of juveniles (Minshull, 1978).

Clarias gariepinus is a common and important fish throughout Zimbabwe (Bell-Cross, 1976) but it may not be as successful in lakes as it is in rivers

---

* Subsequent to writing, Trewavas (1981) has proposed that this cichlid be re-named *Oreochromis (Ny.) macrochir* (Boulenger); however, the name *Sarotherodon macrochir* has been retained in this paper for convenience. – Ed.

(Marshall, 1977). This may be caused by a lack of suitable fish food (Bruton, 1976) and large specimens in Lake McIlwaine feed almost entirely on zooplankton (Munro, 1964, 1976; Murray, 1975). The growth and fecundity of this species in the lake has also been described (Clay, 1979).

The only other species examined in detail in Lake McIlwaine is *Alestes imberi* (Marshall and van der Heiden, 1977).

Species composition

The Hunyani River is part of the middle Zambesi River system from which 58 indigenous fish species have been recorded (Bell-Cross, 1972). Twenty-one of these have been recorded from Lake McIlwaine, making a total of 26 when the 5 introduced species are included (see addendum). *Marcusenius rhodesianus* which was described by Maar (1962) from a single specimen taken from the lake is considered to be synonymous with *Marcusenius macrolepidetus* (Bell-Cross, 1976).

Sarotherodon macrochir and *Tilapia rendalli* were introduced in 1956 to improve commercial fishing and reduce weed growth (Lewin, 1957). Both species are now abundant in the lake and are valuable commercial species (Marshall, 1978a). Another African cichlid, *Haplochromis codringtoni*, was introduced from Lake Kariba in 1978 as a snail predator and to improve angling. Some specimens have been taken but it is not yet abundant (D. H. S. Kenmuir, personal communication).

Micropterus salmoides occurs in the Hunyani River above Lake McIlwaine and has reached the lake in small numbers. No clear evidence of its breeding in the lake has been obtained but it appears to have increased in numbers since 1978 (D. H. S. Kenmuir, personal communication). *Cyprinus carpio* reached the lake from fish ponds in the upper Hunyani River catchment. Some were taken in early commercial catches but the last known specimen from Lake McIlwaine was a 22 kg fish found on the shore in 1971. They have probably never bred in the lake and are now very rare or extinct there. In 1973 about 25 Indian carp, *Catla catla*, escaped from a pond into the lake but have never been found since and probably failed to establish themselves.

The predatory cichlid, *Serranochromis robustus*, is now widely distributed throughout Zimbabwe (Toots and Bowmaker, 1976) and occurs in Lake Robertson immediately below Lake McIlwaine. It has not yet been recorded from the latter but as it has been deliberately (but often illegally) introduced into many water bodies in Zimbabwe it could eventually reach Lake McIlwaine.

Population changes and abundance

Commercial fishing data can be used to provide an assessment of population changes of the 5 most important species in the Lake (Fig. 21). Since the smallest commercial nets used in the lake have a 75 mm stretched mesh, smaller fish are under-represented but some indications of population changes in these species can be gained from other information.

Some species have always been rare as Lake McIlwaine is probably at or near the limit of their distributions. These include *Hippopotamyrus discorhynchus* which is rarely taken, and *Eutropius depressrostris* which has only been recorded once.

The mottled eel, *Anguilla nebulosa labiata*, is occasionally taken by anglers or on long lines and the largest specimen weighed 8.28 kg (Bell-Cross, 1976). The numbers of these species may decline because of the closure of Cabora Bassa Dam on the lower Zambesi in 1975 as this presents a formidable obstacle to the upstream migration of elvers in the Zambesi system. However, since some at least are able to surmount the Kariba dam wall (Balon, 1975) they may also be able to cross Cabora Bassa and may not disappear completely from this part of the Zambesi system.

Several cyprinids have been unable to adapt to lacustrine conditions as they are typically riverine species (Begg, 1974). *Barbus marquensis* was taken in early commercial catches but has now virtually disappeared from the lake whilst the smaller *Barbus* spp., of which *Barbus paludinosus* is the most widespread, occur where streams flow into the lake (Marshall and Lockett,

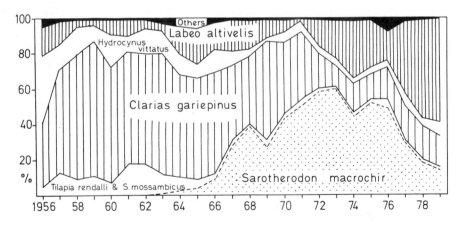

Fig. 21 Composition of the commercial catch from Lake McIlwaine between 1956 and 1979 as a percentage by weight. Data from the Lake McIlwaine Fisheries Research Centre records.

1976). *Barilius zambezensis* and the characid *Micralestes acutidens* are also confined to riverine areas.

Although *Clarias gariepinus* is still an important and common fish in the lake some evidence suggests that it is not as abundant as in the Huyani River (Marshall, 1977). Catches from Lakes McIlwaine and Robertson in 1976 show this clearly (Table 10) and the numbers from the latter were much higher. This lake was formed at the end of 1975 so the *Clarias* taken from there could not have been bred in the lake but must have occurred in the river system. The abundance of *Clarias* in several Zimbabwean lakes, including Lake McIlwaine, appeared to decline after some years (Marshall, 1977) but the reasons for this are unclear. Bruton (1976) has shown that it reaches in maximum size in rivers and suggests that this may be due to a high proportion of fish in their diet. *Clarias* was the major component of the commercial catch until 1966 when it made up 50% by weight of the catch (Fig. 21) but it then declined and by 1979 made up only 15%, a trend also reflected in the actual weight taken (Marshall, 1978a).

Table 10 Mean numbers of *Clarius gariepinus* taken monthly in various nets during 1976 in Lakes McIlwaine and Robertson (after Marshall, 1977). Each net was approximately 25 m long.

Net mesh size (mm)	McIlwaine	Robertson
25	0.18	1.50
38	0.09	8.50
51	0.09	28.12
63	0.34	23.62
76	0.87	2.25
89	1.73	7.62
102	1.54	7.62
114	2.82	6.50
127	0.25	2.62
140	0.14	1.25
152	0.18	0.88
178	—	0.38

Hydrocynus vittatus is the major fish predator in the lake but it is susceptible to gill nets. In the first year of commercial fishing it made up 40% of the catch but later declined to about 10% and, since 1966, has rarely made up more than 5% of the total catch. In Lake Kariba it makes up about 5 to 10% of the total catch; the higher figure being increased through the availability as food of the sardine, *Limnothrissa miodon* (Junor and Marshall, 1979).

The most striking changes in the lake occurred amongst the cichlids (Fig. 21). The indigenous *Sarotherodon mossambicus** was never common, even in the early years (Lewin, 1957) and is now rarely encountered. *Tilapia rendalli* increased until 1962 when it made up nearly 20% of the total catch. Marginal vegetation, especially *Nymphea* beds, was extensive during this period (Munro, 1966) and their destruction has been attributed to *Tilapia rendalli* (Junor, 1969). Blue-green algae may also have played a part, however, as the loss of the macrophytes coincided with the increasing intensity of algal blooms. *Tilapia rendalli* now makes up about 2% of the commercial catch but this does not reflect its true abundance as it is able to avoid being caught in gill nets.

Tilapia sparrmanii was recorded by Munro (1966) but no others were found until 1975 (Marshall and Lockett, 1976). This species is usually associated with quiet water and aquatic vegetation where it can avoid the Tigerfish, *Hydrocynus vittatus*, (Jackson, 1961; Bell-Cross, 1976) and it may well have declined since the loss of the *Nymphea* beds in Lake McIlwaine. *Pseudocrenilabrus philander* is also associated with weeds in the lake and occurs in shallow, sheltered areas. *Haplochromis darlingi* is common in shallow areas and is a major component of the population in these areas (Marshall and Lockett, 1976). It is occasionally taken in commercial catches but it forms a large proportion of the subsistence anglers' catch (Marshall, 1978a).

Sarotherodon macrochir was introduced in 1956 and did not appear in commercial catches until 1963. It rapidly became the most important species in the lake (Marshall and Lockett, 1976; Marshall 1978a) and by 1972 it made up 60% of the commercial catch. This increase occurred after permanent algal blooms developed in the lake and *Sarotherodon macrochir* was found to feed extensively on blue-green algae (Minshull, 1978). Since *Sarotherodon niloticus* was able to digest these plants (Moriarty, 1973) it is possible that *Sarotherodon macrochir* may also be able to do so.

Labeo altivelis has shown a steady increase since 1970 and made up 60% of the total commercial catch in 1979. The reasons for this are not clear but they may be due to the increasingly successful effluent control programme. Blue-green algae have decreased and periphyton appears to have increased as a result of the deeper light penetrations. This would be beneficial to *Labeo altivelis* and possibly detrimental to *Sarotherodon macrochir* although this species also feeds on epiphytes.

---

* Trewavas (1981) has proposed the name *Oreochromis (O.) mossambicus* (Peters) for this cichlid. – Ed.

Table 11 Species composition at 5 poisoning stations in Lake McIlwaine. Numbers refer to the number of times each species occurred at each station expressed as a percentage (after Marshall and Lockett, 1976)

Species	Tiger Bay	Crocodile Creek	Research Bay	Pelican Harbour	Carolina Bank
Sarotherodon macrochir	100	100	100	100	100
Tilapia rendalli	—	40	—	—	30
Haplochromis darlingi	100	100	100	100	100
Pseudocrenilabrus philander	80	60	100	60	100
Alestes imberi	60	20	80	40	30
Micralestes acutidens	—	—	20	—	—
Hydrocynus vittatus	60	80	80	60	30
Labeo altivelis	80	60	80	100	60
Labeo cylindricus	60	20	40	60	60
Barbus paludinosus	100	100	100	100	100
Barbus trimaculatus	40	60	20	20	30
Barbus lineomaculatus	—	40	40	20	100
Barbus radiatus	40	40	60	20	30
Marcusenius macrolepidotus	60	20	—	20	30
Clarius gariepinus	100	100	80	100	100
Micropterus salmoides	—	—	—	20	—

During the course of a poisoning programme Marshall and Lockett (1976) were able to make estimates of the abundance of fish in shallow water (< 2 m deep). They found 17 species in shallow water (Table 11) of which only 6 made up the bulk of the population. These were *Sarotherodon macrochir, Tilapia rendalli, Haplochromis darlingi, Labeo altivelis, Barbus paludinosus* and *Clarias gariepinus* (Fig. 22). Some species that were under-represented include *Hydrocynus vittatus* and *Alestes imberi* which live in deeper water and feed near the surface, and *Marcusenius macrolepidetus* which feeds on benthic fauna and may also occur in deeper water. The solitary occurrence of *Micropterus salmoides* does not provide evidence of breeding as this fish may have escaped from nearby ponds.

Some estimates of standing crop were made but these vary widely (Table 12); this was attributed to fish movements in response to water temperature (Marshall and Lockett, 1976). The estimates of standing crop obtained refer to small areas of shallow water only and thus cannot be applied to the lake as a whole. No reliable estimates of total standing crop in Lake McIlwaine have yet been obtained.

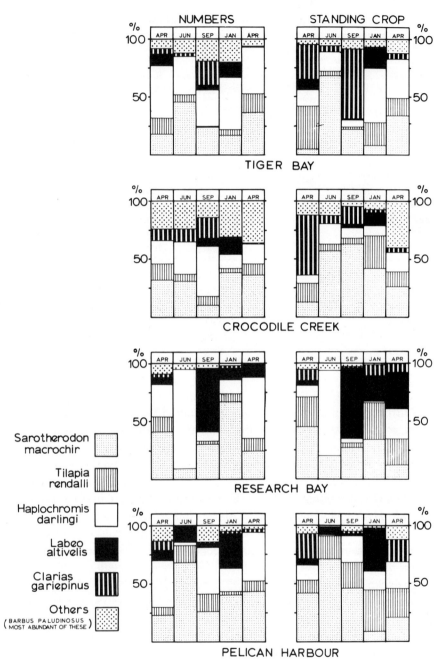

Fig. 22 The composition by numbers and standing crop of juvenile fish as a percentage at four shallow water stations (Marshall and Lockett, 1977).

Table 12 Standing crops at five shallow stations (after Marshall and Lockett, 1976); estimates given in g m^{-2}.

Station	April 1974	June 1974	September 1974	January 1975	April 1975
Tiger Bay	33.2	1.3	28.3	64.1	4.3
Crocodile Creek	5.7	1.8	5.2	58.5	17.8
Research Bay	24.8	0.1	116.4	48.0	27.2
Carolina Bank	19.8	4.0	8.3	—	—
Pelican Harbour	23.4	93.6	6.2	34.3	27.7

Breeding biology

Few detailed studies of breeding biology have been made in Lake McIlwaine, the exceptions being those on *Alestes imberi* and *Sarotherodon macrochir* (Marshall and Van der Heiden, 1977; Marshall, 1978b) whilst some data on *Clarias gariepinus* are also available (Clay, 1979). Most fish in the lake are potamodrometic and cannot breed until the rivers flow and they can migrate upstream.

This is illustrated clearly by *Alestes imberi* which has a short breeding season from about November to February (Fig. 23). This is the period when river flows are the greatest (see B. R. Ballinger and J. A. Thornton, this volume) and although no definite evidence of upstream migration by *Alestes imberi* was obtained from Lake McIlwaine (Marshall and Van der Heiden, 1977) it is known to do so in the Mwenda River flowing into Lake Kariba (Bowmaker, 1973).

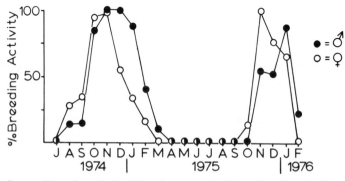

Fig. 23 Seasonal breeding activity of *Alestes imberi* (after Marshall and Van der Heiden, 1977).

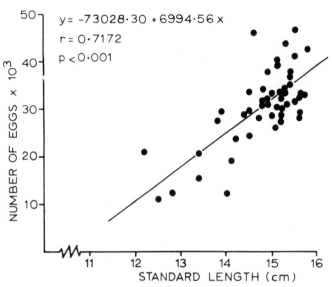

Fig. 24 Eggs produced by *Alestes imberi* in relation to standard length ($n = 54$) (after Marshall and Van der Heiden, 1977).

Many fish which migrate upstream to breed are 'total spawners', i.e., spawning only once during a short breeding season, and these are associated with very high fecundity (Lowe McConnell, 1975). This is shown clearly by *Alestes imberi* where a 16 cm female could produce 50,000 eggs although 30,000 to 40,000 eggs is more typical (Fig. 24). The Lake McIlwaine population appeared to be more fecund than others, as Crass (1964) noted a Natal specimen with only 14,000 eggs whilst Lake Kariba specimens contained 5000 to 20,000 eggs (Bowmaker, 1973).

Clarias gariepinus was found to migrate upstream in the Mwenda River (Bowmaker, 1973) but in general it spawns amongst inundated aquatic or semi-aquatic plants on a lake or river edge (Holl, 1968; Van der Waal, 1972; Bruton, 1979; Willoughby and Tweddle, 1979). Potamodromesis is not therefore invariable in this species and in Lake McIlwaine it could exhibit both types of breeding behaviour. It appears to have a longer breeding season as a result and males appear to maintain a degree of breeding activity throughout the year (Fig. 25).

By contrast the cichlids are not potamodrometic and their breeding behaviour is characterised by nest building and a high level of parental care (Fryer and Iles, 1972). Male *Sarotherodon macrochir* excavate nests in 'arenas' for display and courtship whilst the females brood the eggs and young in their mouths. The breeding season is longer than it is in potamodrometic species (Fig. 26), and they are able to breed several times each season. The main breeding period is just before the rainy season but a rise in lake level can stimulate breeding activity as new nests have been noted on newly flooded land within 12 hours of a water level rise (Marshall, 1978a).

The fecundity of cichlids is relatively low as parental care reduces mortality of eggs and young. *Sarotherodon macrochir* ovaries contain from 500 to 1000 eggs and the females have a high brooding efficiency (Fig. 27). The number of eggs produced was similar to other tilapias (Lowe McConnell, 1955; Cridland, 1961; Welcomme, 1961; Siddiqui, 1977) but the brooding efficiency was generally higher than that recorded for *Sarotherodon leucosticta* (Welcomme, 1967). This suggests that fertility, as distinct from fecundity, is high in *Sarotherodon macrochir* and may be a factor contributing to its success in Lake McIlwaine (Marshall, 1978a).

Food and feeding relationships

In a detailed study of the food of Lake McIlwaine fish, Munro (1967) was able to investigate five species in-depth and make observations on six less im-

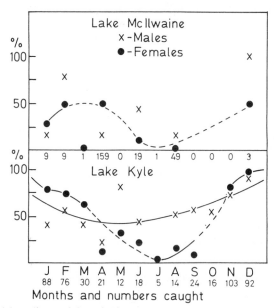

Fig. 25 Seasonal breeding activity of *Clarias gariepinus* as a percentage of fish showing breeding activity. Data from Kyle Dam, Zimbabwe, show the breeding seasons more clearly (re-drawn from Clay, 1979).

165

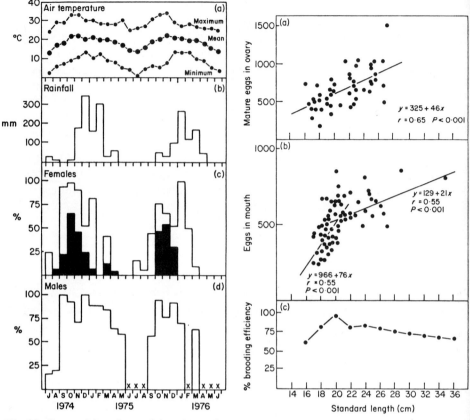

Fig. 26 Seasonal breeding activity of *Sarotherodon macrochir* in relation to temperature (a), and rainfall (b). (c) Percentage of females (open histogram) and ripe females (solid histogram) and (d) males of > 20 cm S.L. taken. x = < 10 fish taken (after Marshall, 1979).

Fig. 27 Fecundity of *Sarotherodon macrochir*; (a) eggs in ovaries in relation to standard length (b) eggs mouth-brooded in relation to standard length, and (c) brooding efficiency in relation to standard length (after Marshall, 1979).

portant ones. He also noted that large *Clarias gariepinus* fed primarily on zooplankton, an aspect studied in greater depth by Murray (1975). *Tilapia rendalli* and *Sarotherodon macrochir* were investigated by Caulton (1975) and Minshull (1978) respectively, whilst the diet of the small characid *Alestes imberi* has also been described (Marshall and Van der Heiden, 1977). Using data available from other sourthern African lakes a fairly comprehensive outline of feeding relationships can thus be obtained.

Mormyridae
In general the mormyrids are adapted for benthic feeding, although some are predatory to some extent (Petr, 1968; Bowmaker, 1973b; Joubert, 1975; Blake, 1977). *Marcusenius macrolepidotus* is the most abundant mormyrid in Lake McIlwaine and feeds mainly on benthic insects (Table 13) of which chironomid larvae and pupae were most abundant. *Mormyrus longirostris* is restricted in the lake and is relatively uncommon. In Lake Kariba it fed on a wide variety of benthic insects (Bowmaker, 1973b; Joubert, 1975) and it undoubtedly takes these in Lake McIlwaine although no investigations have been made. The mormyrids tend to include the same groups of prey organisms in their diets and their trophic relationships are obscure (Corbet, 1961; Blake, 1977). In Lake McIlwaine the two species are probably separated by habitat as *Marcusenius macrolepidotus* appears throughout the lake whilst *Mormyrus longirostris* occurs only in rocky areas.

Table 13 The percentage composition by volume of the diet of *Marcusenius macrolepidotus* (after Munro, 1967)

Ephemeroptera:	*Povilla adusta* nymphus	6.4%
Coleoptera:	*Berosus* sp. larvae	1.1
Odonata:	Coenagriidae nymphus	3.0
	Libellulidae nymphus	5.1
	Aeshnidae nymphus	3.0
Diptera:	*Chaoborus* sp. larvae & pupae	0.4
	Chironomus sp. larvae	1.9
	Chironomidae & Tanypodinae larvae	69.2
	Chironomid pupae	9.6
Other items		0.3

Characidae
The most important predatory fish in the lake is the Tigerfish, *Hydrocynus vittatus*, which exerts a major influence on fish populations in central and southern Africa (Jackson, 1961). A total of 462 *Hydrocynus vittatus* from Lake McIlwaine were examined by Munro (1967) and data for fish 20.1 to 40.0 cm in length showed them to be almost entirely piscivorous (Table 14). No small specimens (< 20 cm) were examined but these probably also feed on fish as in Lake Kariba piscivory began at only 4 cm in length (Kenmuir, 1975) whilst *Hydrocynus forskalli* took fish at this length in Lake Chad (Lauzanne, 1975).

The behaviour of small fish, especially cichlids is governed by the need to

Table 14 The diet of *Hydrocynus vittatus* in Lake McIlwaine given as percentage composition by volume (after Munro, 1967) for two size classes

Food item	20.1–30.0 cm	30.1–40.0 cm
Aquatic insects	7.61	6.64
Terrestrial insects	15.12	4.05
Fish: *Tilapia* spp.	20.97	13.94
Haplochromis darlingi	22.97	31.54
Barbus spp.	5.30	—
unidentified cichlids	27.92	43.83
Amphibia	0.43	—

escape predation by *Hydrocynus vittatus* and they shoal in shallow 'nursery areas' (Donnelly, 1969). As they become larger their vulnerability decreases and they can move into deeper water. In Lake McIlwaine the mean prey length for all sizes of *Hydrocynus vittatus* levelled off at about 6 cm (Fig. 28). The effect of this is shown by the behaviour of juvenile *Sarotherodon macrochir* which remain in very shallow water until they reach a length of about 7 cm (Marshall and Lockett, 1976) but move into deeper water after reaching 10 cm in length (Fig. 29). *Haplochromis darlingi* also remain in very shallow water until they reach about 10 cm in length, but *Tilapia rendalli* stays longer as it feeds on plant detritus in shallow water.

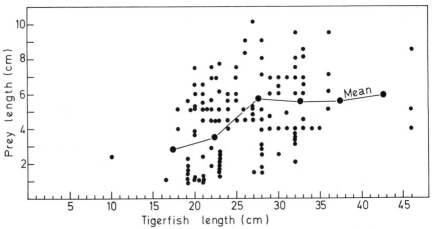

Fig. 28 The relationship between *Hydrocynus vittatus* predator length and cichlid prey length (re-drawn from Munro, 1967).

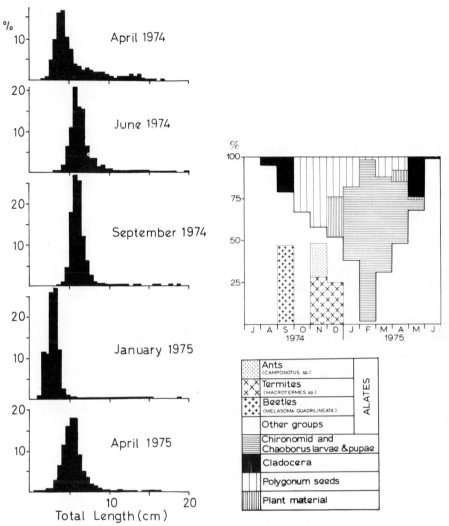

Fig. 29 Length frequencies of *Sarotherodon macrochir* in shallow water showing migration to deeper water after 10 cm length is reached (after Marshall and Lockett, 1976).

Fig. 30 Seasonal variation of food taken by *Alestes imberi* (n = 480) (after Marshall and Van der Heiden, 1977).

In the period from 1956 to 1965 predation by *Hydrocynus vittatus* appeared to have a significant effect on the tilapias; Munro (1967) estimated that each Tigerfish could consume 160 tilapias per year and considered that a reduction of the population might be desirable. Commercial fishing does appear to have reduced their numbers (Fig. 21) whilst the tilapias have increased considerably. Analysis of later commercial catches shows no relationship between *Hydrocynus vittatus* and tilapia catches and the predator probably now has a negligible effect on their recruitment.

The other common characid is the wide ranging *Alestes imberi* which is an omnivorous species frequently taking food off the surface. Insect alates were most important from April to December (Fig. 30), with a wide variety of insects being taken. The most abundant were small cicadellid leaf hoppers but aquatic and terrestrial Diptera, Hymenoptera, Coleoptera, Orthoptera, Ephemeroptera and Hemiptera were also taken (Marshall and Van der Heiden, 1977). They also took advantage of seasonal emergences of insects such as termites, ants and beetles.

Polygonum seeds were important from October to January and other plant material included grass which was most frequently taken in December following a rise in lake level. Chironomids and *Chaoborus* are most abundant in the lake from January to April (Marshall, 1978c) and were an extremely important food item during this period. In February they formed over 95% of the diet but this declined until they made up only 5% in May.

A similar diet was noted by Munro (1967) but he found that *Povilla adusta* was an important component. This species is no longer abundant in the lake (Marshall, 1978c) probably because of the reduced quantity of macrophytes which were abundant in 1962–63 (Munro, 1966). Similar results were reported from Lake Kariba where some *Alestes imberi* were found to take fish (Donnelly, 1971; Bowmaker, 1973) although none did so in Lake McIlwaine (Marshall and Van der Heiden, 1977).

Cyprinidae
Labeo altivelis is the most important cyprinid in the lake but no detailed stomach analyses have been undertaken. They are known to feed on periphyton and Muno (1967) noted the occurrence of algal filaments, diatoms and plant fibres as well as Bdelloid-type Rotifera and Cladocera. In Lake Kariba Bowmaker (1973) suggested that it depended on the bottom muds both as a substrate and directly as food. Burne (1971) found sand in their gut contents and suggested that they feed on sandy bottoms and that diatoms formed the bulk of their food. They may also be able to utilise *Microcystis* as *Labeo rohite* was able to do so under laboratory conditions (Ahmad, 1967).

The only other cyprinid gut analyses from Lake McIlwaine show that *Barbus paludinosus* feeds on diatoms, small chironomid larvae, Cladocera, Copepoda, Rotifera and filamentous algae. Three juveniles (< 3 cm) each contained about 100 *Chydorus globosus* (Munro, 1967).

Clariidae

Clarias gariepinus is an omnivorous species which can play an important role as a fish predator under certain circumstances (Bruton, 1979). Its predatory role in Lake McIlwaine is limited but it feeds on a variety of aquatic organisms (Table 15). A wide range of food organisms were taken by this species with dipterous larvae and pupae (mainly chironomids) and zooplankton being the most important items. A striking feature is the change from mainly Diptera in smaller fish to zooplankton in larger ones (Fig. 31).

Table 15 The food of *Clarias gariepinus* in Lake McIlwaine expressed as percentage composition by volume (after Munro, 1967)

Food item	20–40 cm	40–60 cm	> 60 cm
Plant material	+	0.2	0.5
Cladocera & Copepoda	0.4	20.9	65.5
Odonata nymphus	2.9	1.9	0.8
Other aquatic insects	1.8	0.7	0.3
Dipterous larvae & pupae	75.4	45.2	14.0
Terrestrial insects	8.0	10.3	+
Molluscs	8.8	18.0	8.8
Other items	—	0.9	9.3
Fish	2.8	1.9	0.9

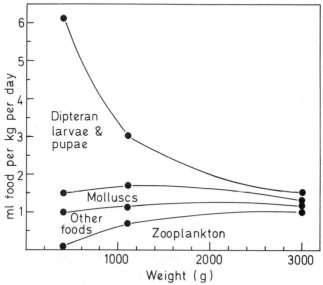

Fig. 31 Apparent changes in the diet of *Clarias gariepinus* in relation to increasing size (re-dawn from Munro, 1967).

This shift to plankton feeding has been noted by several authors (Schoonbee, 1969; Bowmaker, 1973b; Jackson, 1961b) and an attempt to explain why *Clarias gariepinus* smaller than 40 cm total length do not feed on zooplankton was made by Murray (1975). He examined the efficiency of filter feeding by gillrakers and found that the straining surface area increased isomorphically with fish length, whilst the space between the gillrakers increased linearly. His data did not explain why smaller *Clarias gariepinus* did not utilise zooplankton although they should have been able to do so and small captive specimens readily took zooplankton (also noted by Bruton, 1979).

Dietary differences were probably related to the different habitats as smaller fish were more common in shallow water where chironomid larvae and pupae were most abundant (Marshall, 1978c). Large fish occurred in open water and were frequently seen feeding on the surface (Marshall, unpublished) in the manner described by Bruton (1979). Some evidence suggests that *Clarias gariepinus* requires fish in its diet to reach maximum size and that it grows larger in rivers where these are more readily available (Bruton, 1976). This appears to have occurred in Lake McIlwaine where *Clarias gariepinus* are considerably smaller than those from the Hunyani River downstream (Marshall, 1977).

Cichlidae

The cichlids that occur in the lake tend to be catholic in their diet but with some degree of specialisation. *Haplochromis darlingi* is common in shallow water (Marshall and Lockett, 1977) but few data on their dietary habits are available. Munro (1967) examined 131 stomachs and found that chironomid larvae and pupae were mainly taken. Other insects and occasionally gastropods were also taken, whilst fry fed mainly on Cladocera. *Haplochromis codringtoni* feeds mainly on gastropods (Mitchell and Gahamadze, 1976) and was introduced to the lake in an attempt to reduce snail populations and the attendant problems of bilharzia and liver fluke infection. This fish has not yet become well established and there is no evidence of snail control.

Sarotherodon macrochir is the most abundant cichlid in the lake and the most specialised feeder being primarily microphytophagous and is able to take phytoplankton as well as epiphytic and benthic algae. Minshull (1978) examined the stomachs of fry and juveniles which fed mainly on diatoms and *Microcystis aeruginosa* (Table 16). They appeared to be unselective, however, with little difference between their stomach contents and the available food. Munro (1967) examined 10 juveniles which had all fed on diatoms with Cladocera and filamentous algae as other important components.

Munro (1967) found that adult *Sarotherodon macrochir* took Cladocera,

Table 16 The abundance of food organisms on the shallow littoral substrate and in the stomachs of *Sarotherodon macrochir* fry and juveniles expressed as a percentage by numbers (after Minshull, 1978).

Food item	Substrate	Fry	Juveniles
Diatoms	49	60	47
Microcystis aeruginosa	42	28	47
Other algae	7	8	5
Plant detritus	1	1	1
Zooplankton	1	1	–
Fish larvae	–	1*	–

* Probably ingested during capture; there is no other evidence to suggest that *Sarotherodon macrochir* may be piscivorous.

diatoms, filamentous algae and some *Microcystis*. Fish collected more recently however were found to contain large quantities of *Microcystis* and *Anabaena* as well as diatoms (Marshall, unpublished). During periods of blue-green algae blooms the flesh of *Sarotherodon macrochir* was tainted and unpalatable and this may have been caused by *Anabaena* which was ingested at the time. *Sarotherodon macrochir* have also been observed scraping epiphytes from plant stems and this could be an important component of their diet (Marshall, unpublished).

Early work suggested that cichlids could not digest blue-green algae (Fish, 1955) but it has since been shown that *Sarotherodon niloticus* is able to do so (Moriarty, 1973). It seems likely that *Sarotherodon macrochir* can also do this (Minshull, 1978) and this is possibly the main factor behind its success in the lake. The increase in this species coincided with the appearance of the major algal blooms (Fig. 21).

The food of 170 *Sarotherodon mossambicus* was examined by Munro (1967) and higher plants, filamentous algae and diatoms were most important (Table 17). This species is now rare in the lake, possibly because of competi-

Table 17 The food of *Sarotherodon mossambicus* in Lake McIlwaine as a percentage by volume (after Munro, 1967)

Higher plants (including *Lagarosiphon major*)	31.0%
Filamentous algae and diatoms	52.1
Dipteran larvae and pupae	5.2
Cladocera and copepods	11.7

tion with *Sarotherodon macrochir* and *Tilapia rendalli* which take the same food organisms.

Tilapia rendalli is a versatile species that normally feeds on submerged macrophytes (Junor, 1969; Munro, 1967; Caulton, 1975, 1977). However, this is an opportunistic species and Munro (1967) found that chironomids were taken during the rainy season from November to March (Fig. 32). This is the period when these insects are most abundant in the lake (Munro, 1966; Marshall, 1978c) and illustrated the flexible dietary habits of *Tilapia rendalli*. Munro (1967) found that *Lagarosiphon* and *Nymphea* comprised 60% of the food of *Tilapia rendalli* over a year and this may have contributed to the decline of these plants in the lake (Junor, 1969). The loss of these plants is reflected in the diets of juvenile and sub-adult *Tilapia rendalli* which were examined by Caulton (1975).

Some juvenile fish examined by Munro (1967) fed principally on Cladocera, diatoms and filamentous algae; fish < 5 cm also took Rotifera, insect eggs and small chironomid larvae. By contrast Caulton (1975) found that very few benthic or planktonic animals were taken and that diatoms, colonial blue-green algae and macrophytic plants or detritus were the major food items. Varying quantities of blue-green algae (*Microcystis* spp.) were also taken and during the periods of peak phytoplankton blooms in May and June the young fish fed almost exclusively on these algae.

Sub-adult *Tilapia rendalli* also fed heavily on cichlid fry. This suggested adverse conditions when little plant material was available and is another example of the opportunistic feeding of these species, which took advantage

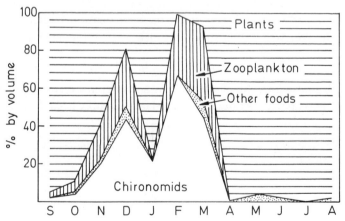

Fig. 32 Monthly variations in the percentage composition by volume of the diet of *Tilapia rendalli* (re-drawn from Munro, 1967).

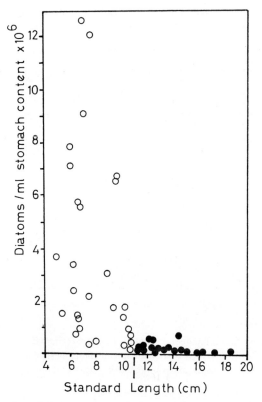

Fig. 33 The proportion of diatoms present in the stomach of *Tilapia rendalli* captured during November 1974; juveniles = 0, and sub-adults = ● (after Caulton, 1975).

of the abundance of fry in November and December (Caulton, 1975). This has been recorded in other habitats lacking in submerged macrophytes (Junor, 1969; Minshull, 1969).

Caulton (1975) also noted a change in the diet of juvenile and sub-adult *Tilapia rendalli* (Fig. 33). The former consumed a greater proportion of diatoms with about 59% by weight of plant macro-fragments. These fragments were typical of the marginal allochthonous detritus derived from the post-summer die-off of the emergent plant, *Polygonum senegalense*. Diatoms played a smaller role in the diet of sub-adult fish with about 80% macro-fragments. These appeared to have been stripped from living *Polygonum*.

A comparison was made between *Tilapia rendalli* from Lake McIlwaine and those from nearby Cleveland Dam which had an abundant aquatic vege-

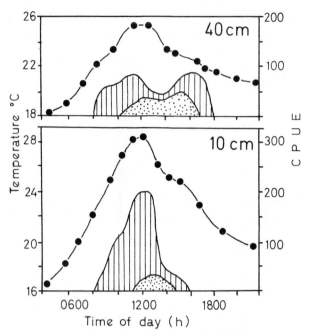

Fig. 34 Daily inshore and offshore movements of *Tilapia rendalli* during October 1974; juveniles are shown as vertical shaded portions, and sub-adults as shaded portions. CPUE is the catch per unit effort (●) at various times during the day (after Caulton, 1975).

tation and where fish were able to feed on 'preferred' food. The most striking difference between the two populations was that the Cleveland Dam fish had larger amounts of food in an apparently full stomach than did the McIlwaine ones. Juveniles of both populations appeared to feed at the same level of intensity but sub-adults from Lake McIlwaine seemed unable to reach the same level of repleteness as the Cleveland Dam fish. This suggested that food availability in Lake McIlwaine might be limiting to sub-adult or larger juvenile fish. Unless the food eaten was of higher quality or more easily digested then Lake McIlwaine fish could be expected to grow more slowly than Cleveland Dam ones. Unfortunately no comparative growth data are available.

Diurnal movements

Daily movements into and away from warm shallow waters is characteristic of many cichlid fishes in African waters (Welcomme, 1964; Donnelly, 1969;

Bruton and Boltt, 1975: Caulton, 1975). This may be a predator avoidance tactic since it is most evident in immature fish (Fryer, 1961; Jackson, 1961a; Donnelly, 1969) but it also occurs in the absence of predators (Welcomme, 1964). Caulton (1976, 1978) has proposed that this is a means of improving physiological efficiency and enhancing growth and food utilisation.

This diurnal movement was clearly demonstrated by *Tilapia rendalli* in Lake McIlwaine (Fig. 34) where both juvenile and sub-adult fish moved into

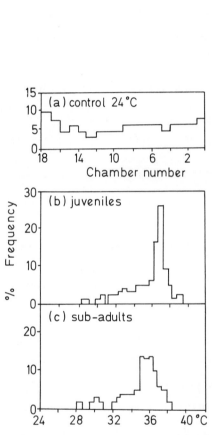

Fig. 35 Temperature selection by *Tilapia rendalli* in a thermal gradient test-tank where the temperature was uniform at 24°C (a) and graduated between 24 and 40°C; juveniles (b) and sub-adult (c) populations are shown (after Caulton, 1975).

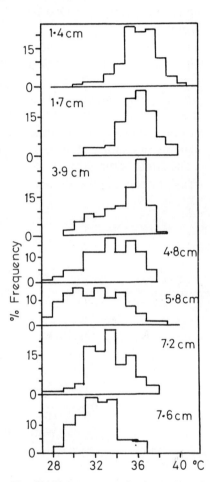

Fig. 36 Temperature selection by juvenile *Sarotherodon macrochir* in relation to fish length (after Minshull, 1978).

shallow water as the water temperature increased (Caulton, 1975, 1976). Fish first appeared in water 40 cm deep at 0800 when the water temperature reached 22°C. They first appeared in water 10 cm deep at 0930 and 1000 when the temperature exceeded 23.5°C. A corresponding decline in juvenile numbers at 40 cm took place at this time. Water temperatures decreased after 1200 and the fish moved back into deeper water. They had all left the shallow water by 1600 (water temperature < 23°C) and the numbers of juveniles at 40 cm increased slightly. By 1700 (water temperature 22°C) there were no fish at 40 cm.

Under experimental conditions juvenile *Tilapia rendalli* selected water temperatures from 31° to 39°C with a mode at 37°C, whilst sub-adults selected temperatures from 32° to 39°C with a mode between 35° and 36.5°C (Fig. 35). Fry selected temperatures as high as 42°C whilst at 24°C there was no selection (Caulton, 1975, 1976). Similar temperature selection occurred with *Sarotherodon macrochir* (Minshull, 1978) and again the smaller fish selected higher temperatures (Fig. 36).

The effect of this strongly thermophilic behaviour was to bring the fish into areas of abundant food supplies during their main feeding period (Fig. 37).

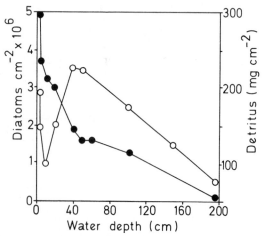

Fig. 37 Distribution of benthic diatoms (o) and organic matter (•) along an offshore transect in the preferred *Tilapia rendalli* habitat in Lake McIlwaine (after Caulton, 1976).

Caulton (1978) discussed the physiological consequences of this and suggested that metabolic efficiency would increase with temperature, to an optimum of about 30°C. He postulated that inshore migrations by juvenile cichlids might be a means of ensuring more efficient growth rates.

Table 18 Cichlid abundance on a gentle gradient beach recorded with a cast net (after Minshull, 1978); numbers caught per cast are shown

Time	September 1977			December 1977		
	°C (5 cm)	S. macrochir	T. rendalli	°C (5 cm)	S. macrochir	T. rendalli
0600	16.5	–	–	20.6	4.0	3.0
0900	20.7	–	–	23.0	10.7	6.7
1200	24.7	–	–	29.3	13.6	4.7
1500	24.3	–	–	27.9	0.5	3.0
1800	19.1	–	–	26.3	6.2	3.4
2100	17.6	0.8	1.8	23.3	–	–
2400	17.0	1.0	0.5	22.0	–	–
0300	15.5	1.4	1.1	21.2	3.0	4.0

The pattern of movement does not appear to be constant, however, and Minshull (1978) noted seasonal differences (Table 18). During the colder months it appeared that cichlids moved into shallow water at night. A normal pattern was recorded in December after water temperatures had risen. No explanation for this was put forward but it may be that no metabolic benefits would be gained until water temperatures approached 30°C. Thus, this migration might not occur during the cold months and there could be some other reason for the inshore movement at night. Even this can vary as Marshall and Lockett (1976) showed a significant increase in shallow water standing crop, brought about by inshore movements, during June when water temperatures did not exceed 20°C.

Trophic relationships

From the data on food habits of the major species some indication of trophic relationships can be gained (Fig. 38). Most food niches in the lake would appear to be well utilised, with the possible exception of the oligochaetes. These are abundant in the lake (Marshall, 1971, 1978c; see B. E. Marshall, this volume) and are dominated by *Branchiura sowerbyi* Beddard. This animal inhabits soft sediments and is probably able to avoid predation by withdrawing several centimetres into the mud. They may also have been overlooked in fish stomachs as they fragment easily.

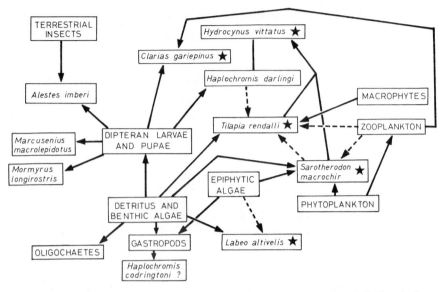

Fig. 38 Simplified food web for the major fish species in Lake McIlwaine. Solid lines indicate the most important, and broken lines the less important, relationships. Commercially important species are indicated by an asterisk (adapted from Marshall, 1978c).

Fish production

The production of fish from the lake is of considerable importance and an attempt to assess production was made by Marshall (1978a). The main sources of fish are from the commercial fishery, angling (recreational and subsistence) and illegal netting.

The commercial fishery began in 1956 and uses gill nets with a minimum mesh size of 75 mm with about 2000 to 2500 m being set each night. The main fish taken are *Sarotherodon macrochir*, *Labeo altivelis*, *Clarias gariepinus* and *Hydrocynus vittatus*, and the total catch for the period 1972 to 1976 averaged 110 tonnes per year (Table 19). Catches increased substantially from 1968 onwards when *Sarotherodon macrochir* became the most important species (Fig. 39). The maximum catch was recorded in 1973 when 140 tonnes were taken. From 1956 to 1968 *Clarias gariepinus* was the most important species followed by *Sarotherodon macrochir* until 1978 when *Labeo altivelis* became most important (Fig. 21). This reflects the trophic status of the lake and suggests that improved water quality favours *Labeo altivelis*.

Angling data was collected by an angling survey which estimated that recreational anglers took about 62 tonnes per year whilst subsistence anglers

Table 19 Annual fish landings in tonnes by the Lake McIlwaine commercial fishery (after Marshall, 1978a)

Species	1972	1973	1974	1975	1976	Mean
Sarotherodon macrochir	60.4	83.7	43.8	59.1	42.5	58.9
Tilapia rendalli	3.0	1.4	1.8	2.4	4.5	2.6
Labeo altivelis	15.7	31.9	34.1	30.7	13.7	25.2
Hydrocynus vittatus	4.1	6.1	3.4	5.1	3.3	4.4
Clarius gariepinus	21.3	16.1	21.2	15.6	15.5	17.9
Others	0.7	0.5	1.9	1.7	7.8	2.5
Totals	105.2	139.7	106.2	114.6	87.3	110.5

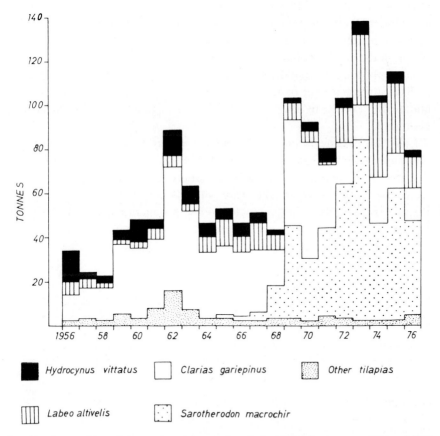

Fig. 39 Composition of the commercial catch from Lake McIlwaine between 1956 and 1976; data from 1977 onwards are thought to be questionable and may not give a true reflection of the total catch (after Marshall, 1978a).

took about 52 tonnes per year. The total of 114 tonnes per year is comparable to the commercial catch but this is probably an underestimate caused by the difficulties of collecting angling information at Lake McIlwaine.

The relationship between commercial and angling catches (Table 20) shows that, although there are considerable differences, most common species are taken by the various fishing methods. Subsistence anglers take large numbers of small cichlids which accounts for the high proportion of *Haplochromis darlingi* in their catches. Both angling methods take *Tilapia rendalli* which is not important in the commercial catches, whilst the commercial fishing takes large numbers of *Labeo altivelis*.

The quantities of fish taken illegally cannot be estimated accurately but are believed to be at least 50 to 75 tonnes per year (Marshall, 1978a). This is probably a considerable underestimate as illegal fishing has increased since 1976 and may now remove well over 100 tonnes per year. Most illegal fishermen use beach seine netting methods and this may be having a detrimental effect on the cichlid populations.

Total production from the lake, therefore, can be conservatively estimated at 300 tonnes per year or 120 kg ha^{-1} yr^{-1}. This represents a landed value of about Z\$ 180,000 in 1978 and is clearly a significant factor in the local economy.

Lake McIlwaine is one of the most productive lakes in Africa with a photosynthetic rate of 1.64 to 6.03 g C m^{-2} h^{-1} (Robarts, 1979; see R. D. Robarts, this volume) and its annual fish production is comparable to that from many other African lakes (Fig. 40). The strong oxycline in Lake McIlwaine is probably a limiting factor to fish production as the shallow Lake George, Uganda, produces 179 kg ha^{-1} yr^{-1} (Fryer and Iles, 1972) whilst

Table 20 Species composition as a percentage by numbers of the commercial and angling catches at Lake McIlwaine (after Marshall, 1978a)

Species	Recreational anglers	Subsistence anglers	Commercial fishery
Sarotherodon macrochir	28.7	26.2	56.1
Tilapia rendalli	17.7	35.2	2.8
Haplochromis darlingi	4.9	29.4	0.1
Labeo altivelis	0.6	5.0	28.3
Clarias gariepinus	8.1	0.1	5.8
Marcusenius macrolepidotus	–	–	2.6
Hydrocynus vittatus	0.3	–	3.9
Alestes imberi	30.2	4.0	0.2

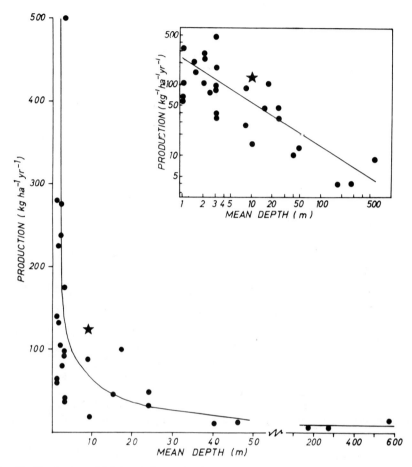

Fig. 40 Production of fish in relation to depth in African lakes; asterisk shows the position of Lake McIlwaine. Inset shows the same data plotted on a logarithmic scale (re-drawn from Fryer and Illies, 1972; after Marshall, 1978a).

having a photosynthetic rate of 0.5 to 0.8 g C m^{-2} h^{-1} (Ganf, 1975).

Fish production from Lake McIlwaine is clearly linked to its eutrophication (Fig. 41). There is therefore some concern that high production cannot be maintained as the nutrient content of the lake declines. Since light is now the main factor limiting primary production (Robarts and Southall, 1977; Robarts, 1979; R. D. Robarts, this volume) a decrease in the blue-green algae population could be followed by enhanced growth of epiphytic and benthic algae. This could account for the recent increase in *Labeo altivelis* at the expense of *Sarotherodon macrochir* and total production could be maintained at a high level, although with a changed species composition.

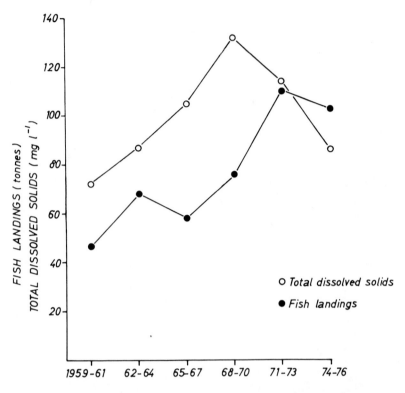

Fig. 41 The relationship between total dissolved solids and fish catches in Lake McIlwaine. The mean annual data for three-year periods are shown (after Marshall, 1978a).

Acknowledgement

This paper is published with the approval of the Director of National Parks and Wild Life Management, Zimbabwe.

References

Ahmad, M. R., 1967. Observations on the effect of feeding *Labeo rohita* (Ham.) with *Microcystis aeruginosa* Kutz. Hydrobiol., 29: 388–392.

Balon, E. K., 1975. The eels of Lake Kariba: distribution, taxonomic status, growth and density. J. Fish. Biol., 7: 797–815.

Begg, G. W., 1974. The distribution of fish of riverine origin in relation to the limnological characteristics of the five basins of Lake Kariba. Hydrobiol., 44: 277–286.

Bell-Cross, G., 1972. Fish fauna of the Zambezi River system. Arnoldia Rhod., 5: 1–9.

Bell-Cross, G., 1976. The fishes of Rhodesia. National Museums and Monuments Pub., Salisbury.

Addendum

Fish species recorded from Lake McIlwaine. Common names used are the 'Standard' names proposed by Jackson (1975) and used in Bell-Cross (1976). The names used locally differ considerably in many cases.

1. Anguillidae
    Anguilla nebulosa labiata Peters, 1852. African mottled eel.

2. Mormyridae
    Hippopotamyrus discorhynchus (Peters, 1852). Zambesi parrotfish.
    Marcusenius macrolepidotus (Peters, 1852). Bulldog.
    Mormyrus longirostris Peters, 1952. Eastern bottlenose.

3. Characidae
    Alestes imberi Peters, 1852. Imberi.
    Hydrocynus vittatus Castelnau, 1861. Tigerfish.
    Micralestes acutidens (Peters, 1852). Silver robber.

4. Cyprinidae
    Cyprinus carpio Linnaeus, 1758. Common carp.*
    Barbus lineomaculatus Boulenger, 1903. Line-spotted barb.
    Barbus marquensis A. Smith, 1841. Largescale yellowfish.
    Barbus paludinosus Peters, 1852. Straightfin barb.
    Barbus radiatus Peters, 1853. Beira barb.
    Barbus trimaculatus Peters, 1852. Threespot barb.
    Barilius zambezensis (Peters, 1852). Barred minnow.
    Labeo altivelis Peters, 1852. Hunyani labeo.
    Labeo cylindricus Peters, 1852. Redeye labeo.

5. Schilbeidae
    Eutropius depressirostris (Peters, 1852). Butter catfish.

6. Clariidae
    Clarius gariepinus (Burchell, 1822). Sharptooth catfish.

7. Centrarchidae
    Micropterus salmoides (Lacepede, 1802). Largemouth bass.*

8. Cichlidae
    Haplochromis codringtoni (Boulenger, 1908). Green happy.*
    Haplochromis darlingi (Boulenger, 1911). Zambesi happy.
    Pseudocrenilabrus philander (M. Weber, 1897). Southern mouthbrooder.
    Sarotherodon macrochir (Boulenger, 1912). Greenhead tilapia.*
    Sarotherodon mossambicus (Peters, 1852). Mozambique tilapia.
    Tilapia rendalli (Boulenger, 1896). Redbreast tilapia.*
    Tilapia sparrmanii (A. Smith, 1840). Banded tilapia.

* Introduced species.

Blake, B. F., 1977. The effect of the impoundment of Lake Kainji, Nigeria, on the indigenous species of mormyrid fishes. Freshwat. Biol., 7: 37–42.

Bowmaker, A. P., 1973. An hydrobiological study of the Mwenda River and its mouth, Lake Kariba. Ph.D. Diss., University of the Witwatersrand.

Bruton, M. N., 1976. On the size reached by *Clarias gariepinus*. J. Limnol. Soc. Sth. Afr., 2: 57–58.

Bruton, M. N., 1979a. The breeding biology and early development of *Clarias gariepinus* (Pisces: Clariidae) in Lake Sibaya, South Africa, with a review of breeding in the subgenus *Clarias (Clarias)*. Trans. Zool. Soc. Lond., 35: 1-45.

Bruton, M. N., 1979b. The food and feeding behaviour of *Clarias gariepinus* (Pisces: Clariidae) in Sibaya, South Africa, with emphasis as a predator of cichlids. Trans. Zool. Soc. Lond., 35: 47–114.

Bruton, M. N. and R. E. Boltt, 1975. Aspects of the biology of *Tilapia mossambica* Peters (Pisces: Cichlidae) in a natural freshwater lake (Lake Sibaya, South Africa). J. Fish. Biol., 7: 423–445.

Burne, R. H., 1971. A comparative study of the diets of *Labeo congoro, Labeo altivelis* and *Tilapia mossambica* in the Sanyati Basin of Lake Kariba. Lake Fisheries Research Institute Project Rep. No. 12.

Caulton, M. S., 1975. Diurnal movement and temperature selection by juvenile and sub-adult *Tilapia rendalli* Boulenger (Cichlidae). Trans. Rhod. Scient. Ass., 56: 51-56.

Caulton, M. S., 1976. The energetics of metabolism, feeding and growth of sub-adult *Tilapia rendalli* Boulenger. D.Phil. Diss., University of Rhodesia.

Caulton, M. S., 1978. The importance of habitat temperatures for growth in the tropical cichlid *Tilapia rendalli* Boulenger. J. Fish. Biol., 13: 99–112.

Clay, D., 1979. Sexual maturity and fecundity of the African catfish (*Clarias gariepinus*) with an observation on the spawning behaviour of the Nile catfish (*Clarias lazera*). Zool. J. Linn. Soc., 65: 351–365.

Crass, R. S., 1964. Freshwater fishes of Natal. Shuter and Shooter, Pietermaritzburg.

Cridland, C. C., 1961. Laboratory experiments on the growth of *Tilapia* spp. The reproduction of *Tilapia esculenta* under artificial conditions. Hydrobiol., 18: 177–184.

Corbet, P. S., 1961. The food of non-cichlid fishes in the Lake Victoria Basin, with remarks on their evolution and adaptation to lacustrine conditions. Proc. Zool. Soc. Lond., 136: 1–100.

Donnelly, B. G., 1969. A preliminary survey of *Tilapia* nurseries on Lake Kariba during 1967/68. Hydrobiol., 34: 195–206.

Fish, G. R., 1951. Digestion in *Tilapia esculenta*. Nature Lond., 167: 900.

Fryer, G., 1961. Observations on the biology of the cichlid fish *Tilapia variabilis* Boulenger in the northern waters of Lake Victoria (East Africa). Rev. Zool. Bot. Afr., 64: 1–33.

Fryer, G. and T. D. Iles, 1972. The cichlid fishes of the Great Lakes of Africa. Oliver and Boyd, London.

Ganf, G. G., 1975. Photosynthetic production and irradiance–photosynthesis relationships of the phytoplankton from a shallow equatorial lake (Lake George, Uganda). Oecologia (Berl., 18: 165–183.

Holl, E. A., 1968. Notes on the spawning behaviour of barbel *Clarias gariepinus* Burchell in Rhodesia. Zool. Afr., 3: 185–188.

Jackson, P. B. N., 1961a. The fishes of Northern Rhodesia. Government Printer, Lusaka.

Jackson, P. B. N., 1961b. The impact of predation, especially by the Tigerfish (*Hydrocynus vittatus* Castelnau) on African freshwater fishes. Proc. Zool. Soc. Lond., 136: 603–662.

Jackson, P. B. N., 1975. Common and scientific names of the fishes of southern Africa. Part 11. J. L. B. Smith Institute of Ichthyology Spec. Pub. No. 14, Rhodes University, Grahamstown.

Junor, F. J. R., 1969. *Tilapia melanopleura* Dum. in artifical lakes and dams in Rhodesia with special reference to its undesirable effects. Rhod. J. Agric. Res., 7: 61–69.

Junor, F. J. R. and B. E. Marshall, 1979. The relationship between the Tigerfish and the Tanganyika sardine in Lake Kariba. Rhod. Sci. News, 13: 111–112.

Joubert, C. S. W., 1975. The food and feeding habits of *Mormyrus deliciosus* (Leach, 1818) and *Mormyrus longirostris* Peters, 1852 (Pisces: Mormyridae, in Lake Kariba. Kariba Studies, 5: 68–85.

Kenmuir, D. H. S., 1973. The ecology of the Tigerfish *Hydrocynus vittatus* Castelnau in Lake Kariba. Occ. Pap. Nat. Mus. Rhod. B., 5: 115–170.

Kenmuir, D. H. S., 1975. The diet of fingerling Tigerfish *Hydrocynus vittatus* Castelnau in Lake Kariba. Arnoldia (Rhod.), 9 (7): 1–8.

Lauzanne, L., 1975. Régimes alimentaires d'*Hydrocyon forsakalii* (Pisces: Characidae) dans le Lac Tchad et ses tributairs. Cah. O.R.S.T.O.M., sér. Hydrobiol., 9: 105–121.

Lewin, G., 1957. Commercial fishing on Lake McIlwaine. Proc. First Fish. Day S. Rhod., Government Printer, Salisbury.

Lowe McConnell, R. H., 1975. Fish communities in tropical freshwaters. Longman, London.

Maar, A. A., 1962. *Marcusenius smithers*: sp. nov. and *Gnathonemus rhodesianus*: sp. nov. (Mormyridae) from the Zambesi River system and *Barbus hondeensis*: sp. nov. (Cyprinidae) from the Pungwe River. Occ. Pap. Nat. Mus. S. Rhod., 3: 780–784.

Marshall, B. E., 1977. On the status of *Clarias gariepinus* (Burchell) in large man-made lakes in Rhodesia. J. Limnol. Soc. Sth. Afr., 3: 67–68.

Marshall, B. E., 1978a. An assessment of fish production in an African man-made lake (Lake McIlwaine, Rhodesia). Trans. Rhod. Scient. Ass., 59: 12–21.

Marshall, B. E., 1978b. Aspects of the ecology of benthic fauna in Lake McIlwaine, Rhodesia. Freshwat. Biol., 8: 241–249.

Marshall, B. E., 1978c. Lake McIlwaine after twenty-five years. Rhod. Sci. News, 12: 79–82.

Marshall, B. E., 1979. Observations on the breeding biology of *Sarotherodon macrochir* (Boulenger) in Lake McIlwaine, Rhodesia. J. Fish. Biol., 14: 419–424.

Marshall, B. E. and C. A. Lockett. 1976. Juvenile fish populations in the marginal areas of Lake Kariba, Rhodesia. J. Limnol. Soc. Sth. Afr., 2: 37–42.

Marshall, B. E. and J. T. van der Heiden, 1977. The biology of *Alestes imberi* Peters (Pisces: Characidae) in Lake McIlwaine, Rhodesia. Zool. Afr., 12: 329–346.

Minshull, J. L., 1969. An introduction to the food web of Lake Sibaya, Northern Zululand. Newslett. Limnol. Soc. Sth. Afr., 13 (Suppl.): 20–25.

Minshull, J. L., 1978. A preliminary investigation of the ecology of juvenile *Sarotherodon macrochir* (Boulenger) on a shallow shoreline in Lake McIlwaine, Rhodesia. M.Sc. Thesis, University of Rhodesia.

Mitchell, S. A. and C. Gahamadze, 1976. *Haplochromis codringtoni* (Cichlidae) and bilharzia on Lake Kariba. Lake Kariba Fisheries Research Institute Project Rep. No. 25.

Moriarty, D. J. W., 1973. The physiology of digestion of blue-green algae in the cichlid fish, *Tilapia nilotica*. J. Zool. Lond., 171: 25–39.

Munro, J. L., 1964. Feeding relationships and production of fish in a Southern Rhodesian lake. Ph.D. Diss., University of London.

Munro, J. L., 1966. A limnological survey of Lake McIlwaine, Rhodesia. Hydrobiol., 28: 281–308.

Munro, J. L., 1967. The food of a community of East African freshwater fishes. J. Zool. Lond., 151: 389–415.

Murray, J. L., 1975. Selection of zooplankton by *Clarias gariepinus* (Burchell) in Lake McIlwaine, a eutrophic Rhodesian reservoir. M.Sc. Thesis, University of Rhodesia.

Robarts, R. D., 1979. Underwater light pnetration, chlorophyll *a* and primary production in a tropical African lake (Lake McIlwaine, Rhodesia). Arch. Hydrobiol., 86: 423–444.

Robarts, R. D. and G. C. Southall, 1977. Nutrient limitation of phytoplankton growth in seven tropical man-made lakes, with special reference to Lake McIlwaine, Rhodesia. Arch. Hydrobiol., 79: 1–35.

Schoonbee, H. J., 1969. Notes on the food habits in Lake Barberspan, Western Transvaal, South Africa. Verh. Int. Verein. Limnol., 17: 689–701.

Siddiqui, A. Q., 1977. Reproductive biology, length-weight relationship and relative condition of *Tilapia lencosticta* (Trewavas) in Lake Naivasha, Kenya. J. Fish. Biol., 10: 251–260.

Stewart, L. H., 1957. Fisheries management in National Parks waters. Proc. First Fish. Day. S. Rhod. Government Printer, Salisbury.

Toots, H. and A. P. Bowmaker, 1976. *Serranochromis robustus jallae* (Boulenger, 1896) (Pisces: Cichlidae) in a Rhodesian highveld dam. Arnoldia (Rhod.), 7 (39): 1–16.

Trewavas, E., 1981. Nomenclature of the tilapias of southern Africa. J. Limnol. Soc. Sth. Afr., 7: 42.

Van der Waal, B. C. W., 1974. Observations on the breeding habits of *Clarias gariepinus* (Burchell). J. Fish. Biol., 6: 23–27.

Willoughby, N. G. and D. Tweddle, 1978. The ecology of the catfish *Clarias gariepinus* and *Clarias ngamensis* in the Shire Valley, Malaŵi. J. Zool. Lond., 186: 507–534.

Welcomme, R. L., 1964. The habits and habitat preferences of the young of Lake Victoria Tilapia (Pisces: Cichlidae). Rev. Zool. Bot. Afr., 70: 1–28.

Welcomme, R. L., 1967. The relationship between fecundity and fertility in the mouth-brooding cichlid fish *Tilapia leucosticta*. J. Zool. Lond., 151: 453–468.

Avifauna of Lake McIlwaine
M. J. F. Jarvis

A number of bird counts have been made on Lake McIlwaine and its shoreline. This paper does not attempt a comprehensive checklist but rather it selects some species and traces changes in their abundance as the lake vegetation and nutrient levels have altered.

Soon after the dam was completed in 1952, extensive aquatic vegetation developed including reed beds along portions of the shore and on submerged termite mounds, water lily beds (*Nymphea* spp.) and large areas of water hyacinth (*Eichhornia crassipes*) (see M. J. F. Jarvis *et al.*, this volume). The reeds were mainly *Phragmites* and *Typha*, and initially *Typha* appeared to be most abundant.

As the appropriate authorities intended Lake McIlwaine to serve as a recreational and boating area as well as a water supply for Salisbury City residents, the aquatic vegetation conflicted with this aim. Consequently chemical and manual methods were used to reduced the amount of water hyacinth on the lake. The chemical used was 2,4-D amine which kills a variety of broad-leaf plants. Consequently with the reduction in *Eichhornia crassipes* there was a similar reduction in *Nymphea* and other aquatic vegetation. Associated with these changes were a number of changes in avifauna.

Changes in the avifauna

In a summary of bird surveys undertaken at the lake up to 1971 (Table 21) total numbers and average counts are not given due to the variety of sample methods used. However, if the percentage of surveys recording each species are compared, a meaningful picture emerges. The situation after 1971 (Table 22) is compared with the data from Table 21 (Table 23).

It appears that a number of species increased in percentage occurrence after 1971 when the aquatic vegetation had gone. These included the Egyptian Goose (*Alopochen aegyptiacus*), Fulvous Duck (*Dendrocygna bicolor*), Hottentot Teal (*Anas hottentota*), Knob-bill Duck (*Sarkidiornis melanotos*), Red-eye Pochard (*Netta erythrophthalma*), Red-bill Teal (*Anas erythrorhyncha*), Spurwing Goose (*Plectropterus gambensis*), White-face Duck (*Dendrocygna viduate*) and Red-knobbed Coot (*Fulica cristata*). All these species except the Coot, Egyptian Goose and Pochard use the dam mainly as a loafing area and range out to farm dams to feed. To a certain extent the Pochard also ranges out but they do seem to find a considerable amount of food within the dam. The Coot and Egyptian Goose have learnt to utilise shoreline vegetation and the Egyptian Goose is the only duck species that regularly nests on the lake.

The species that have declined since 1971 are all birds that utilise floating vegetation for food, either directly or by feeding on life forms in the vegetation. These species included the Pygmy Goose (*Nettapus auritus*), Whiteback Duck (*Thalassornis leuconotus*), Dabchick (*Tachybaptus ruficollis*), Purple Gallinule (*Porphyrio porphyrio*), Lesser Gallinule (*Porphyrio alleni*), Moorhen (*Gallinula chloropus*) and African Jacana (*Actophilornis africanus*).

Other avifauna

Other species recorded for Lake McIlwaine include wading birds (Tree, 1974, 1976, 1977), but no attempt has been made to show detailed changes in the

Table 21 Summary of bird surveys up to the end of 1971 showing data source and percentage occurrence of each species

Species	Boulton (1960)		Anon. (1963)		Boulton & Woodall (1965)		Conway (1971)		Vernon (1970)		Total counts	Percentage occurrence
	Poss. obs.	Total obs.	Poss. obs.	Total obs.	Poss. obs.	Total obs.	Poss. obs.	Total obs.	Poss. obs.	Total obs.		
Black Duck	2	2	3	5	0	7	2	3	28	274	291	12
Egyptian Goose	1	1	5	5	1	7	3	3	97	274	291	37
Fulvous Duck	1	2	1	5	0	7	3	3	19	274	291	8
Hottentot Teal	1	2	1	5	1	7	3	3	32	274	291	13
Knob-bill Duck	2	2	2	5	3	7	2	3	56	274	291	22
Moccoa Duck	0	2	0	5	0	7	0	3	2	274	291	1
Red-eye Pochard	2	2	5	5	5	7	3	3	90	274	291	36
Pigmy Goose	1	2	5	5	6	7	1	3	85	274	291	34
Red-bill Teal	2	2	4	5	3	7	3	3	80	274	291	32
Spurwing Goose	0	2	0	5	0	7	3	3	7	274	291	3
White-back Duck	2	2	5	5	1	7	3	3	64	274	291	26
White-face Duck	1	2	4	5	7	7	3	3	101	274	291	40
Dabchick	2	2	5	5	3	7	–	–	–	–	17	59
Pruple Gallinule	2	2	5	5	7	7	–	–	–	–	17	71
Lesser Gallinule	1	2	3	5	0	7	–	–	–	–	17	24
Moorhen	2	2	5	5	7	7	–	–	–	–	17	82
Red-knobed Coot	2	2	5	5	3	7	–	–	–	–	17	59
African Jacana	2	2	5	5	7	7	–	–	–	–	17	82

Table 22 Summary of bird surveys after December 1971 showing data source and percentage occurrence of each species

Species	Russel & Tree (1973)		Jarvis (1975)		National Waterfowl Survey (1978)		Total counts	Percentage occurrence
	Poss. obs.	Total obs.	Poss. obs.	Total obs.	Poss. obs.	Total obs.		
Black Duck	1	14	1	2	6	45	61	13
Egyptian Goose	14	14	2	2	32	45	61	79
Fulvous Duck	3	14	0	2	9	45	61	20
Hottentot Teal	10	14	0	2	14	45	61	39
Knob-bill Duck	11	14	2	2	21	45	61	56
Maccoa Duck	0	14	0	2	0	45	61	0
Red-eye Pochard	14	14	1	2	29	45	61	92
Pygmy Goose	0	14	0	2	2	45	61	3
Red-bill Teal	12	14	2	2	44	45	61	95
Spurwing Goose	14	14	2	2	19	45	61	57
White-back Duck	0	14	0	2	4	45	61	7
White-face Duck	14	14	2	2	23	45	61	64
Dabchick	5	14	2	2	–	–	16	44
Purple Gallinule	0	14	0	2	–	–	16	0
Lesser Gallinule	0	14	0	2	–	–	16	0
Moorhen	0	14	1	2	–	–	16	6
Red-knobed Coot	9	14	2	2	–	–	16	69
African Jacana	8	14	2	2	–	–	16	63

Table 23 A comparison of the percentage occurrence prior to December 1971 and subsequently

Species	Percentage occurrence		Status after 1971		
	Before 1971	After 1971	Increased	Decreased	Static
Black Duck	12	13			×
Egyptian Goose	37	79	×		
Fulvous Duck	8	20	×		
Hottentot Teal	13	39	×		
Knob-bill Duck	22	56	×		
Maccoa Duck	1	0			×
Red-eye Pochard	36	92	×		
Pygmy Goose	34	3		×	
Red-bill Teal	32	95	×		
Spurwing Goose	3	57	×		
White-back Duck	26	7		×	
White-face Duck	40	64	×		
Dabchick	59	44		×	
Purple Gallinule	71	0		×	
Lesser Gallinule	24	0		×	
Moorhen	82	6		×	
Red-knobbed Coot	59	69	×		
African Jacana	82	63		×	

Table 24 A comparison of duck counts at Lake McIlwaine with those for the whole of Zimbabwe based on National Waterfowl Survey results from 1972–78 (after Jarvis, in press); means include zero counts

Species	Lake McIlwaine data		National data	
	Mean No. per count	Relative abundance	Mean No. per count	Relative abundance
Black Duck	0.6	9	0.3	7
Egyptian Goose	115.0	2	2.9	5
Fulvous Duck	4.0	8	0.1	11
Hottentot Teal	1.6	7	0.1	9
Knob-bill Duck	19.0	5	3.3	3
Maccoa Duck	0.0	12	0.1	12
Red-eye Pochard	114.0	3	1.5	4
Pygmy Goose	0.1	11	0.1	10
Red-bill Teal	227.0	1	7.8	1
Spurwing Goose	8.4	6	0.4	8
White-back Duck	1.3	10	0.4	6
White-face Duck	104.0	4	6.5	2

status or relative abundance of these species except for some Anatidae (Table 24). this is compared with the overall relative abundance of duck in the whole country (Jarvis, in press).

Comparing the order of relative abundance of Anatidae shows that the four most abundant species at Lake McIlwaine are the same as the nationally most abundant. These are the Red-bill Teal, Egyptian Goose, Red-eye Pochard, White-face Duck and Knob-bill Duck. Of the remaining seven species, Lake McIlwaine has relatively more Spurwing Geese, Hottentot Teal and Fulvous Duck but relatively fewer Black Duck, White-back Duck and Pygmy Geese.

Some rarer birds have been recorded at the lake. For example Garganey (*Anas querquedula*) (A. J. Tree, personal communication), Cape Teal (*Anas capensis*) in flocks of up to about 10 (Jarvis and Tree, unpublished) and Cape Shoveller (*Anas smithii*) (A. J. Tree, personal communication; Jarvis, 1976).

Periodically when water levels are low large flocks of Openbill Stork (*Anastomus lamelligerus*) arrive to feed on stranded lamellibranchs (Tree, 1973; Van der Heiden, 1973; Henwood, 1973). Low water levels and exposed mud banks also attract many migrant wading birds (Tree, 1974). Flamingos are periodic visitors when the water is low. Records include Boulton and Woodall (1970), Tree (1977a) and Borrett (1969). In 1969 several sightings were made of a Gull-billed Tern (*Gelochelidon nilotica*) (Campbell, 1969; Manson, 1969) and in 1978 sightings were made of a Slaty Egret (*Egretta vinaeceigula*) (Evans, 1979).

Discussion

Observed changes in frequency of occurrence of some waterfowl species at Lake McIlwaine seem to be linked with changes in aquatic vegetation. The destruction of aquatic vegetation has been attributed to the influence of the herbivorous fish, *Tilapia melanopleura* (Junor, 1969; see B. E. Marshall, this volume). Although this fish can severely damage aquatic vegetation, it appears unlikely that it was the prime cause of the vegetation change in Lake McIlwaine. This appears to be the widespread use of the herbicide 2,4-D.

The elimination of floating vegetation must also have resulted in a large reduction in snail and other small life forms, thus reducing available food for several bird species. Since light penetration and wave action would also have increased, this probably produced changes in the planktonic flora and fauna which have been noted elsewhere in this volume.

Changes to the ecosystem due to 2,4-D amine may be more extensive than realised at first since some indication exists (Tinker, 1971) that other life forms, including freshwater lamellibranchs, might be adversely affected. Marshall (1975) showed that some lamellibranch species were apparently absent from the lake in 1973 whereas they were abundant in 1962–63, and although it is likely that drought and water level fluctuations produced these anomalies it could be worth considering the possible effects of herbicide application.

Acknowledgements

This analysis was made possible by the unpublished notes and reports of observers mentioned in the text. Lake McIlwaine falls within the McIlwaine Recreational Park and reports by staff of the Department of National Parks and Wild Life Management form a considerable proportion of the data used. This data is published with the authority of the Director of National Parks and Wild Life Mangement, Zimbabwe.

References

Borrett, R., 1969. Flamingos at Lake McIlwaine. The Honeyguide, 59: 37.
Boulton, R., 1961. Atlantica Field Notes No. 1. Atlantica Ecological Research Station, Salisbury.
Boulton, R. and P. Woodall, 1970. Fundamentals of field ornithology. Rhod. Ornithol. Soc. Pub.
Campbell, N. A., 1969. Gull-billed Tern (?) at Lake McIlwaine. The Honeyguide, 59: 30.
Evans, P. J. 1979. Letter to the Editor. The Honeyguide, 97: 37.
Henwood, P., 1973. Openbill Storks at Lake McIlwaine. The Honeyguide, 75: 29.
Jarvis, M. J. F., 1976. Cape Shoveller at Lake McIlwaine. The Honeyguide, 86: 43.
Jarvis, M. J. F., in press. Distribution and abundance of waterfowl (Anatidae) in Zimbabwe. Proc. Fifth Pan-African Ornithol. Congr., Malawi, 1980.
Junor, F. J. F., 1969. *Tilapia melanopleura* Dum. in artificial lakes and dams in Rhodesia, with special reference to its undesirable effects. Rhod. J. Agric. Res., 7: 61–69.
Manson, A. J. and C. Manson, 1969. Gull-billed Tern–further sighting. The Honeyguide, 59: 31.
Marshall, B. E. and A. C. Falconer, 1973. Physico-chemical aspects of Lake McIlwaine (Rhodesia), a eutrophic tropical impoundment. Hydrobiol., 42: 45–62.
Marshall, B. E., 1975. Observations on the freshwater mussels (Lamellibranchia: Unionacea) of Lake McIlwaine, Rhodesia. Arnoldia (Rhod.), 16 (7): 1–15.
Tinker, J., 1971. Unhealthy herbicides. New Scientist, 49: 593.
Tree, A. J., 1973. Birds on Lake McIlwaine. The Honeyguide, 76: 32–35.
Tree, A. J., 1974. Waders in the Salisbury area 1972/74. The Honeyguide, 80: 13–27.
Tree, A. J., 1976. Waders in Central Mashonaland 1974/75. The Honeyguide, 85: 17–27.
Tree, A. J., 1977. Waders in Central Mashonaland 1975–77. The Honeyguide, 92: 21–41.
Tree, A. J., 1977a. Some recent local records of interest. The Honeyguide, 90: 35–37.
Van der Heiden, J. T., 1973. Openbill Storks nesting near Salisbury. The Honeyguide, 76: 23–25.
Vernon, C. J., 1971. Report on the status of Rhodesian waterfowl. Department of National Parks and Wild Life Management Rep., Salisbury.

6 Utilisation, management and conservation

Water pollution: perspectives and control
D. B. Rowe

Although Zimbabwe is an underdeveloped country, water pollution problems have arisen. However, steps have been taken to understand these problems and combat them.

The country lies within the tropics and it seems that many of the findings of research in the temperate zones of the world are not necessarily applicable in the man-made lakes of the tropics. This applies particularly to the problems of eutrophication. Lake McIlwaine, in particular, has received considerable study in this regard and it appears that the diversion of sewage effluent from the lake has been successful in reversing the eutrophication that was occurring. This approach offers one solution to the eutrophication problem and the papers in this volume detail the results of the research effort.

These research results are a tool for the management of the water bodies in the country. Water management in Zimbabwe is under the control of the Division of Water Development of the Ministry of Natural Resources and Water Development. Much of the research effort on freshwaters in the country has been supported and/or commissioned by the Division of Water Development.

This paper is descriptive and provides a background to the problems of water pollution in Zimbabwe. It also describes both the legal and practical approach to water pollution control.

Water pollution

Zimbabwe lies within the tropics. Its average annual rainfall ranges from 1700 mm in the east to 320 mm in the south-west. Most of this rain falls between December and February during the rainy season and hence most of the rivers are non-perennial and cease flowing during the dry season. There is considerable variation from year to year in the run off. The country's economy is based on agriculture and mining, and ninety percent of the population lives in the rural areas. Traditionally drinking water has been drawn from shallow wells and from the rivers, or from pools or sandbeds in the rivers when they cease to flow.

The population is increasing at a rate of 3.6% per annum, increasing from 400,000 in 1900 to over 6 million in 1981. This development has necessitated the provision of additional water resources usually in the form of dams and boreholes. The present position is that most of the major urban centres are supplied from man-made lakes. As many of these centres lie along the central watershed, cities are situated upstream of their sources of water supply. This is true of Salisbury which is located upstream of Lake McIlwaine and hence any wastewater from the city re-enters its source of supply.

In the rural areas the traditional supplies have also been augmented with boreholes and small dams, and water is still carried by the women from these sources to their homes. Although the situation is steadily improving, a large proportion of the rural population relies on untreated surface water for drinking, washing and bathing.

In the 1960s public complaints about water pollution increased considerably due to the expansion of urban, industrial and mining activity. Several lakes were showing signs of eutrophication and in particular Lake McIlwaine was giving cause for concern.

Water pollution control legislation

Prior to 1970 water pollution control legislation was weak. From 1970 water pollution control legislation was improved and has been framed upon the following principles:
a) quality standards should ensure that the waters of the country are preserved in their natural state of purity, that healthy biological life in rivers and lakes is maintained and that where some degree of pollution is unavoidable it is controlled and kept within acceptable limits;
b) discharge standards for wastewater are based upon the quality of the wastewater and not the quality of the receiving water; and,
c) the polluter pays the cost for the control measures.

This current legislation is contained in the Water Act, No. 41 of 1976, and its associated regulations. The important aspects of the Water Act, 1976, are that water pollution is defined, that water pollution is a criminal offence, and that discharge standards are prescribed.

In the Act, water pollution is defined as:

a) such contamination or other alteration of the biological chemical or physical properties of the public stream or water, including changes in the colour, odour, taste, temperature or turbidity; or

b) such discharge of any gaseous, liquid, solid or other substance into the public stream or water; as will, or is likely to create a nuisance or render the public stream or private water, public water or underground water, as the case may be, detrimental, harmful or injurious to the health, safety or welfare of the public or any section thereof, or to any consumer or user of the public water or to any birds, fish or aquatic life, livestock or wild animals.'

It should be noted that this definition covers both nuisance and human health aspects and all animal life.

Legal effect is given to this Act in the provisions of the Water (Effluent and Waste Water Standards) Regulations, 1976, which are appended in full at the end of this section. The Regulations prescribe standards for effluent or wastewater discharged into water and cover pH, temperature, dissolved oxygen content, chemical oxygen demand, oxygen absorbed, undissolved and dissolved solids, numerous toxic substances, detergents and nutrients (nitrogen and phosphorus). The standards are strict and are roughly in line with those of other countries where discharge standards have been laid down. Two sets of standards given in the Regulations; a higher set of standards is laid down for the trout streams in the eastern part of the country (Zone I) and a slightly lower set for the rest of the country (Zone II).

Provision is made in the Act for issuing permits granting exemption from these standards. In practice, such permits are only issued for limited periods whilst pollution control measures are being implemented.

Prior to the promulgation of the Regulations it was common practice for municipalities to treat their sewage in oxidation ponds or biological filters to the standard of 'secondary treated effluent'. The new standards meant that further or tertiary treatment of effluents was required. As physico-chemical methods of treatment proved too expensive, irrigation was generally adopted as the practicable method of treatment (see J. McKendrick, this volume). More recently, a biological method of treatment (the Bardenpho Process) has been developed which is also capable of treating effluents to meet the standards. This process is now being introduced in several centres in Zimbabwe inlcuding Salisbury.

Administration of the Water Act, 1976

The Division of Water Development is responsible for the administration of the Water Act, 1976. The Division investigates reports of water pollution and takes whatever action is necessary. In some instances the pollution can be abated by simple measures; in others the necessary measures are difficult, time-consuming and costly. In all instances, the approach of the Division is a constructive one, of discussing the matter with the polluter and of applying continual pressure on him to formulate proposals for remedial measures and to implement them. Prosecutions are applied only as a last resort if the polluter proves intransigent. Generally, the attitude of industry and municipalities has been co-operative and the approach has worked well.

The disposal of sewage during wet weather

During the long dry season, discharges or spillages seldom occur from well-designed and operated sewage treatment systems in Zimbabwe. However, during the rainy season, significant discharges or spillages do occur from overloaded sewage reticulation systems and sewage works, and from saturated irrigation lands if the effluent is used for irrigation. Theoretically, it is possible to prevent such discharges but often only at considerable cost. The problem is one of diminishing returns; a stage is reached where comparatively little pollution remains and extremely costly measures are required to reduce this small amount of pollution even further. Such expenditure may not be warranted if the adverse effect on the environment is negligible. As this is largely the case during the Zimbabwean rainy season, the Division of Water Development has prepared guidelines covering this aspect of water pollution and has recommended the minimum design capacities of both sewers and treatment works related to the dry weather flow (DWF) of the sewage. The design capacities are shown in Tables 1 and 2.

As effluents receiving secondary treatment do not comply with the prescribed standards, they may not be discharged into water courses and normally tertiary treatment is provided by means of irrigation. However, during the rains the irrigated lands may become saturated and the effluent runs off the surface. The Division therefore has accepted that only 1× DWF may be discharged to water courses during prolonged periods of wet weather. If this requirement is met, present indications are that, under Zimbabwean climatic conditions, 90% of the nutrients will be prevented from entering the water courses.

In the case of nitrogen and phosphorus, whilst there may be no adverse effects and animal health risk from discharging them into a river in flood, there may be an accumulation of nutrients in any downstream lakes. Hence, it is the total loads of nutrients entering the river system that are important and the question of how much nitrogen and phosphorus can be discharged to water

Table 1 Recommended sewer capacities

Dry weather flow 10^3 m^3 d^{-1}	Minimum design factor for sewer capacity (multiple of DWF)
0–2	5.25
2–6	4.50
6–20	3.75
20–200	3.00
> 200	2.70

Table 2 Recommended treatment works capacities

Treatment	Minimum design factor for treatment works capacity (multiple of DWF)
At least screening	all flows
At least screening and primary settlement	5 × DWF
Full secondary treatment	3 × DWF

courses during the rains without giving rise to eutrophication problems downstream has been the driving force behind much of the research conducted in Zimbabwe.

Water quality monitoring and research

In order to establish the natural quality of rivers, lakes and groundwaters in Zimbabwe, the Division of Water Development established the National Water Quality Survey in 1976. The aim of the Survey is to provide a firm ecological basis for the multi-purpose management of the water resources of Zimbabwe through a co-ordinated programme of research and monitoring. The Water Quality Monitoring Programme presently consists of a network of 110 stations established on rivers and canals throughout the country, and the programme is being expanded on a phased basis to include further stations on rivers and lakes. Research programmes commissioned by the Division of Water Development through the National Water Quality Survey are usually caried out on a co-operative basis. An example of this is the urban run off investigations presently being carried out on several small catchments in urban areas (see J. A. Thornton and W. K. Nduku, and R. S. Hatherly and K. A. Viewing, this volume), as a combined exercise by the Division, the City of Salisbury, the Institue of Mining Research and the Hydrobiology Research

Unit of the University of Zimbabwe. The Hydrobiology Research Unit has also undertaken research projects to answer specific questions posed by the Division of Water Development, and, supported by grants from the Division and other interested bodies including the City of Salisbury, has been conducting research programmes on Lake McIlwaine since its inception in 1967. Much of this research is summarised in this volume.

Addendum

Prescribed standards of effluent or wastewater as given in the second schedule (Section 3) of the Water (Effluent and Waste Water Standards) Regulations, 1977, promulgated in Rhodesia Government Notice No. 687 of 1977.

Prescribed standards of effluent or wastewater
1. The water shall not contain any colour or have any odour or taste capable of causing pollution.
2. The water shall not contain any radioactive substances capable of causing pollution.
3. The pH of the water shall be, where discharged or disposed of –
   (a) in a Zone I catchment area, between 6.0 and 7.5;
   (b) in a Zone II catchment area, between 6.0 and 9.0.
4. The temperature of the water at the point of discharge shall not exceed –
   (a) in a Zone I catchment area, 25°C;
   (b) in a Zone II catchment area, 35°C.
5. The water shall contain dissolved oxygen to the extent of at least, where discharged or disposed of –
   (a) in a Zone I catchment area, 75 *per centum* saturation;
   (b) in a Zone II catchment area, 60 *per centum* saturation.
6. The chemical oxygen demand of the water, after applying chloride correction, shall not exceed, where discharged or disposed of –
   (a) in a Zone I catchment area, 30 milligrams per litre;
   (b) in a Zone II catchment area, 60 milligrams per litre.
7. The oxygen absorbed by the water shall not exceed, where discharged or disposed of –
   (a) in a Zone I catchment area, 5 milligrams per litre;
   (b) in a Zone II catchment area, 10 milligrams per litre.
8. The total undissolved solids content of the water at the point of discharge shall not be greater than –
   (a) in a Zone I catchment area, 10 milligrams per litre;
   (b) in a Zone II catchment area, 25 milligrams per litre.
9. The total dissolved solids content of the water at the point of discharge shall not –
   (a) in a Zone I catchment area, increase the total dissolved solids content of the receiving water by more than 100 *per centum* and the total dissolved solids content of the effluent shall not exceed 100 milligrams per litre;
   (b) in a Zone II catchment area, exceed 500 milligrams per litre.
10. The water shall not contain soap, oil or grease in quantities greater than, where discharged or disposed of –

(a) in a Zone I catchment area, nil;
(b) in a Zone II catchment area, 2.5 milligrams per litre.

11. The maximum permissible concentrations of chemical constituents permissible in the water which is discharged or disposed of in Zone I or Zone II catchment areas shall be as specified in the following table:

Table
Maximum permissible concentrations of certain chemical constituents

Constituent	Maximum concentration in milligrams per litre	
	Zone I catchment area	Zone II catchment area
Ammonia free and saline (as N)	0.5	0.5
Arsenic (as As)	0.05	0.05
Barium (as Ba)	0.1	0.5
Boron (as B)	0.5	0.5
Cadmium (as Cd)	0.01	0.01
Chlorides (as Cl)	50	100
Chlorine residual (as free chlorine)	nil	0.1
Chromium (as Cr)	0.05	0.05
Copper (as Cu)	0.02	0.5
Cyanides and related compounds (as CN)	0.2	0.2
Detergents (as manoxol-OT)	0.2	1.0
Fluoride (as F)	1.0	1.0
Iron (as Fe)	0.3	0.3
Lead (as Pb)	0.05	0.05
Manganese (as Mn)	0.1	0.1
Mercury (as Hg)	0.5	0.5
Nickel (as Ni)	0.3	0.3
Nitrogen total (as N)	10.0	10.0
Phenolic compounds (as phenol)	0.01	0.1
Phosphates total (as P)	1.0	1.0
Sulphate (as SO_4)	50	200
Sulphide (as S)	0.05	0.2
Zinc (as Zn)	0.3	1.0
Total heavy metals	1.0	2.0

12. The water shall not contain any detectable quantities of pesticide, herbicide or insecticide, nor shall it contain any other substances not referred to elsewhere in these standards, in concentrations which are poisonous or injurious to human, animal, vegetable or aquatic life.

Water supply and sewage treatment in relation to water quality in Lake McIlwaine
J. McKendrick

The population of Greater Salisbury was 313,000 in 1968, but by 1979 the population had risen to 650,000 and the present population is now likely to be close to 1 million. In 1968 the water consumption varied from 63.0 Ml per day during the rainy season to 112.5 Ml per day during the hot, dry season, and in 1979 this had risen to 134.05 Ml per day during the rainy season and 211,28 Ml per day during the hot, dry season. This increase in consumption since 1967–68 is shown in Table 3. To meet this demand, Salisbury obtains its water from a number of dams on the Hunyani River and its tributaries: namely, Prince Edward Dam (3000 Ml), Cleveland Dam (910 Ml), Lake McIlwaine (250,000 Ml), Henry Hallam Dam (9200 Ml) and Lake Robertson (490,000 Ml). Of these, Lake McIlwaine is the main source of water supply to the City.

Table 3 Volume of water pumped to the City of Salisbury between 1967 and 1978 in Ml per year

Year	Volume pumped	Year	Volume pumped
1967–68	34,983.3	1973–74	44,298.8
1968–69	35,771.4	1974–75	55,023.3
1969–70	39,069.2	1975–76	56,898.2
1970–71	43,073.6	1976–77	57,654.2
1971–72	43,322.6	1977–78	60,016.0
1972–73	50,598.0		

The Salisbury urban area drains into the Hunyani River above Lake McIlwaine but below Prince Edward Dam. As a result most of the drainage from the City flows into Lake McIlwaine. While most of the outer sub-urban areas of Greater Salisbury have been developed into 0.4 ha sub-divisions and are served by septic tanks, the whole of the central business area, the industrial sites and the high density residential areas are served by water-borne sewerage reticulation. As with the increased water consumption, the volume of sewage effluent entering the City's sewage treatment plants has increased since 1955 (Table 4).

Prior to 1970 the effluent from these works discharged into the Marimba and Makabusi Rivers which enter the Hunyani River just above Lake McIlwaine. The quality of the effluent discharged from the sewage works was always kept well within any recognised international standards because it has always been

Table 4 Total dry weather flows (DWF) in $m^3\ d^{-1}$ entering the two main sewage treatment works (Crowborough and Firle) since 1955

Year	DWF ($m^3\ d^{-1}$)	Year	DWF ($m^3\ d^{-1}$)
1955	19,090	1967	39,545
1956	17,954	1968	38,045
1957	18,182	1969	40,045
1958	25,454	1970	40,182
1959	26,363	1971	46,136
1960	28,182	1972	52,136
1961	29,091	1973	57,000
1962	33,636	1974	70,097
1963	32,727	1975	79,607
1964	35,454	1976	81,088
1965	34,545	1977	76,481
1966	37,273	1978	86,295

appreciated that high quality effluents were necessary as Zimbabwe is faced with the problem of a seasonal rainfall and is not richly endowed with large perennial rivers. Zimbabwe has, therefore, to rely on storage dams to maintain an adequate water supply all the year round.

Water quality changes in Lake McIlwaine

Water was first drawn from Lake McIlwaine in November 1953 and until 1959 little change was observed in the quality of the raw water. The quality of water in terms of turbidity, colour and planktonic growth, and ease of treatment of Lake McIlwaine water was similar to that obtained from the other two dams (Prince Edward Dam and Cleveland Dam). In common with these dams a seasonal increase in turbidity during the rainy season entailed an increase in alum dosage, which in the case of Lake McIlwaine amounted to a change from an average of 24.0 mg l^{-1} to an average of 35.0 mg l^{-1}. During the period 1954–59 no difficulties were experienced in treating the raw water.

From 1960, periodic algal blooms appeared in the lake and caused purification difficulties at the works. These difficulties, however, were overcome by increasing the alum dosage to 45.0 mg l^{-1} and by the use of flocculant aids.

During mid-1964 there was no evidence to suggest that the similar problems that were expected to arise later in the year would not be overcome with the same treatment. However, when temperatures started to rise in the latter half of the spring (September – October 1964) the algae in Lake McIlwaine multiplied at an exceptionally high rate and produced an unexpectedly intense bloom. The large amount of algae and the resulting high pH of the water (at times up to 9.6) created unusual problems in flocculation and clarification which were extremely difficult to overcome. Following previous practice the alum dosage was increased substantially above 35.0 mg l^{-1} and on one occasion up to 75.0 mg l^{-1}.

With the rise of summer temperatures the demand for water in Salisbury increased rapidly to a weekly average of between 50.6 Ml and 52.8 Ml. Although this amount was just below the rated capacity of the treatment plant, which was 55 Ml at the time, the flocculation and clarification processes became more difficult to control. As large quantities of floc were carried over on to the sand filter beds the loss of head built up extremely rapidly and the filter runs became so short (6 to 8 hours) that the Works' staff had great difficulty in keeping up with the filter backwashing in order to meet the demand. It was noticed furthermore that the stage had been reached when algae were actually passing through the sand filters.

Past experience had shown that the water drawn from the 7.5 m level intake had been the easiest to treat, but in order to overcome the problems encountered in 1964 and in an endeavour to get water from below that polluted with algae, the intake level was changed to the 14 m level. The raw water from that level had an average pH of 7.1 and was free from algae and clear in appearance. The amount of alum required to produce visibly good flocculation with this water was about 32.0 mg l^{-1}. It was noted that the floc was very fine and feathery in appearance and that considerable amounts were still being carried over on to the sand filters. Immediately after the change to the 14 m level it was found that very dirty water was passing through the filters after about 6 to 8 hours and that although the loss of head registered was only 0.5 to 1 m it became necessary to backwash the filters to maintain a clear water. Normally a loss of head of 2 m would be reached before it was necessary to backwash and it was therefore evident that some peculiar circumstances had arisen which allowed the passage of the apparently dirty water, for the alum floc although fine and feathery would be expected to be retained on the sand bed at such a low loss of head.

During the period 1959–64 the 14 m level inlet had been used on occasions in order to avoid the concentration of algae which had been formed with high easterly winds near the inlet tower, but on those occasions no difficulty had been experienced in treating the water. In light of subsequent experience, however, it is most probable that some dirty water was passed into the reticulation but not in sufficient quantity to result in the number of complaints reaching serious proportions.

To ascertain the cause of the dirty water, samples were taken every half hour after backwashing and it was found that the concentration of aluminium in the filtered water was 0.75 mg Al l^{-1}, whereas under normal conditions this figure had previously been under 0.05 mg Al l^{-1}. The concentration of aluminium in the filtered water remained fairly constant for the three hours immediately following backwashing and then rose sharply to 1.5 mg l^{-1} when the water started to appear very dirty. The dirtiness increased to such an extent that after 6 to 8 hours it became necessary to backwash the filters.

A comparison of the aluminium concentrations in the water before and after filtration showed that before filtration most of the aluminium was in solution whereas after filtration most of the aluminium was out of solution. These results suggested that the water leaving the clarifier contained large proportions of unreacted alum, which precipitated out of the water during its passage through the sand in the filters, and this precipitate contributed to the dirty colour of the water. Laboratory tests showed that the addition of 0.125 mg l^{-1} coagulant aid improved the character of the floc to such an extent that it was decided to use it at the Works, still maintaining the draw-off level at 14 m. The results achieved were remarkable as the floc lost its feathery appearance and became larger and heavier, resulting in very little being carried over to the sand filters and the periods between backwashing of the filters were extended in many cases up to 24 hours. Although the water leaving the Works was then apparently clear, the aluminium content remained high and numerous complaints about dirty water were still being received from consumers. Further investigation showed that the manganese concentration of the raw water was 0.5 mg Mn l^{-1} and that of the filtered water 0.4 mg Mn l^{-1}. It thus appeared that the use of coagulant aids prevented the precipitation of the dissolved alum and manganese in the filters but that this precipitation occurred later in the reticulation system, resulting in dirty water being delivered to consumers throughout the City.

During this period, sampling of the water in Lake McIlwaine around the intake tower revealed that there was a heavy concentration of decaying sludge, due, it was thought, to dead algae settling in the river bed near to the intake tower. Because of stratification in the lake the water below the 9 m depth was entirely devoid of oxygen and this effect probably aggravated by the presence of decaying sludge resulted in iron and manganese being released into solution from the underlying rocks and plant material to form complex organic salts. Later tests showed that iron and manganese would go into solution due to stratification without the presence of the sludge. The land under and surrounding the lake is mainly banded ironstone and analyses have shown that both the rock and the vegetation in the area have a very high manganese content (see K. Munzwa, and R. S. Hatherly and K. A. Viewing, this volume).

The lack of oxygen in the water at the 14 m depth prevented the precipitation of the iron and manganese during flocculation and clarification as the period of retention in the Works was insufficient to ensure complete oxidation prior to filtration. It is throught that the complete absence of oxygen may have played a part in preventing the complete formation of the aluminium hydroxide floc. Further aeration occurred, however, during the passage of the water through the sand filters causing some of the iron, manganese and aluminium to come out of solution to give the dirty water appearance in the filter outlet boxes in the period prior to the use of coagulant aids. The disappearance of the dirty water appearance in the filter outlet boxes when coagulant aids were used was probably due to the coagulant aid having an inhibitory or slowing effect on the oxidation processes, thus preventing the precipitation of aluminium and the other compounds in the filters. Whether the subsequent liming, chlorination and natural aeration in the storage reservoirs caused the precipitation to occur later or whether it was the result of the slowing effect of the coagulant aid which retarded the precipitation until after the water had left the filters was not known. Although the apparent clarity of the water after filtration led us to believe that the problem had been overcome, the continued widespread appearance of dirty water throughout the City and the volume of complaints soon disabused us of this optimism.

The difficulties experienced at the 14 m level were in fact far worse than experienced at the 7.5 m level and it was decided of necessity to change back to the latter. The alum dosage was increased to 45.0 mg l^{-1} and although the flocculation appeared to be satisfactory, 0.5 mg Al l^{-1} was still found in the filtered water. It was realised that although the flocculation appeared from visual observations to be satisfactory a true picture was not in fact being obtained as the residual aluminium in the filtered water was still high.

Subsequent laboratory tests involving jar-test flocculation, filtration and residual aluminium estimation showed that over a wide range of alum dosage the results were visually satisfactory. However, the tests indicated quite definitely that minimum residual aluminium would be obtained only at a particular dosage, which at the time of the tests was found to be 55.0 mg l^{-1} of alum. When this dosage was adopted at the Works the final filtered water was found to be free from dissolved aluminium. The apparent anomaly of increasing the alum dosage to reduce the outgoing aluminium content is explained by the fact that the alum is first used for pH correction and the residual alum is then used for floc formation. Insufficient or excess alum will not provide the correct pH for proper floc formation. The return to the 7.5 m level automatically resolved the iron and manganese problems as no iron nor manganese is present above the 7.5 m depth.

The very hard lesson was learnt that one cannot judge flocculation results by visual appearance alone, by jar test or otherwise but that the aluminium content of the filtered water must be checked as part of the routine control of water purification works.

The cause of the problem

Intensive investigations were made at this time to ascertain the cause of this intensive algal bloom in the lake. These investigations revealed that the major contributing factor causing this condition in Lake McIlwaine was the drainage from the Salisbury urban area and in particular the sewage effluents which, although of a high quality, contained high concentrations of nitrogen and phosphorus (Marshall and Falconer, 1973a, 1973b). The drainage had caused rapid eutrophication of Lake McIlwaine resulting in a typical eutrophic lake with algal activity confined exclusively to the epilimnion and a reservoir of available nutreints in the hypolimnion. The normal ecology had been disturbed with the prolific development of blue-green algae, mainly species of *Microcystis* and *Anabaena* (see J. A. Thornton, this volume).

Studies to reduce the pollution of Lake McIlwaine

The foregoing observations indicated that immediate steps had to be taken to reduce the volume of nutrients entering the lake from the Salisbury urban area if existing conditions in the lake were to be improved. It was decided to tackle the problem in two ways. Firstly, to investigate the physico-chemical removal of nutrients on a laboratory scale and pilot plant scale, and secondly to study the use of sewage effluents for crop irrigation as a means of removing the nutrients from the effluents before they found their way back into the natural water courses.

Physico-chemical removal of nutrients
Laboratory studies on the physico-chemical removal of nutrients from raw sewage and sewage effluents were made (McKendrick, 1973). As a result of

Table 5 Analysis of the final effluent obtained by physico-chemical treatment of raw sewage; units mg·l^{-1} except where noted

Constituent	Conc.	Constituent	Conc.
pH (Brönsted Units)	6.8	Colour (Hazen Units)	5
Dissolved oxygen	1.8	C.O.D.	nil
Oxygen absorbed	0.8	Suspended solids	1.2
Soap, oil, grease etc.	nil	Dissolved solids	428.0
Free and saline ammonia	0.1	Arsenic	nil
Total nitrogen	0.3	Chloride	57.0
Total phosphorus	0.018	Total chromium	nil
Detergents (Manoxol O.T.)	0.05	Fluoride	0.08
Iron	0.08	Manganese	0.08
Sulphate	222.9	*E. coli* (No. per 100 ml)	nil

these studies two pilot plants were constructed at one of the sewage works; one to study the physico-chemical treatment of raw sewage and the other to study chemical-biological treatment.

The physico-chemical treatment plant consisted of alum treatment of the incoming raw sewage, followed by clarification, sand filtration, activated carbon filters, ammonia removal using clinoptilolite and chlorination of the final effluent. Analyses of the final effluent from this plant are given in Table 5.

These results showed that this type of treatment produced an effluent which virtually conformed with the World Health Organisation's (W.H.O.) standards for drinking water, and showed the possibility of the use of this type of treatment to augment water supplies which may be necessary in the future

Table 6 Analysis of the final effluent obtained by chemical-biological treatment of raw sewage; units mg l^{-1} except where noted

Constituent	Conc.	Constituent	Conc.
pH (Brönsted Units)	8.0	Colour (Hazen Units)	15
Dissolved oxygen	5.9	C.O.D.	37.5
Oxygen absorbed	5.5	Suspended solids	70.0
Turbidity (Jackson Units)	2.8	Settleable solids (cm^3 l^{-1})	0.8
Free and saline ammonia	2.5	Methylene blue stability	100%
Total nitrogen*	24.45	Chloride	48.0
Total phosphorus	0.6	Total alkalinity (CaCO$_3$)	130.0

* Total nitrogen = nitrate (24.0) + nitrite (0.45).

in Zimbabwe. However, because of the cost involved and the difficulty of importing materials such as activated carbon and clinoptilolite, further studies were discontinued.

The chemical-biological treatment plant consisted of lime treatment of raw sewage followed by clarification, re-carbonation and biological filtration. Analyses of the effluent from the biological filters are shown in Table 6.

The prime object of this part of the project was phosphorus removal accompanied by good nitrification with minimum amortisation of the existing plant. This has been successfully achieved and further research will be carried out on this plant with a view to removing or reducing nitrogen and other constituents to acceptable limits.

Crop irrigation of sewage effluents

It has always been appreciated that sewage effluent and sludge are a valuable source of water and nutrients for agricultural purposes. As one of the major sewage works had 160 ha of arable land, a modest initial programme of land irrigation was embarked upon to study the general removal of nutrients from sewage effluents by land irrigation (Neely, 1975).

A 5 ha plot was developed and planted under lucerne (*Medicago sativa*). This plot was flood irrigated with sewage effluents under controlled conditions for the purpose of this study. Typical results obtained on sewage effluent being applied and the regeneration water are shown in Table 7. Based on the success of these results it was decided to gradually develop the rest of the 160 ha of land, adjusting techniques of irrigation and the types of crops as and when necessary.

These studies revealed that flood irrigation was the most economical means of application of the effluent and that the crops best suited to this type of irrigation were maize, sorghum, sunflower, wheat, lucerne and pasture grasses. However, the constant irrigation of these lands with effluent caused springs and streams to run which are normally dry during the winter dry season. Samples from these and test holes in the lower regions of the lands

Table 7 Analysis of humus tank effluent and regeneration water during the experimental irrigation of lucerne; units mg l^{-1}

Constituent	Humus tank effluent	Regeneration water
Total phosphorus	9–10	0.2
Free and saline ammonia	9–11	trace
Nitrate	4–6	20–25

Table 8 Analysis of sewage effluent and regeneration water during the winter (June to September) and summer (November to December); units mg l^{-1}

Month	Reactive phosphorus	Total phosphorus	Free and saline ammonia	Nitrate	Chloride
June	0.01	0.07	—	trace	18.0
August	0.13	0.93	0.8	nil	25.0
September	0.1	0.26	1.3	nil	20.0
November	2.0	2.71	4.5	nil	102.0
December	2.5	3.73	2.0	2.5	62.0
Sewage effluent	9.0	9.3	12.0	14.0	180.0

were constantly monitored and analytical results from a typical sampling point are shown in Table 8. These results showed that during the dry season the effluent is draining into the soil, giving up its nutrients, mixing with the sub soil water and thus raising the water table forming springs and streams. In the rainy season, however, the soil is saturated and sewage effluent mixed with rainwater tended to run off the surface. These preliminary studies showed, therefore, that irrigation of sewage effluents for nutrient removal was successful during the dry season, but only partially so during the rainy season.

Although the emphasis was on the primary need to remove nutrients from the effluent, it was realised that this could only be achieved if good farming practices were followed. As a result, land had to be kept dry at certain times of the year so that ploughing, ripening and reaping could be carried out, which meant that a large area of expensive land was required. The rainfall problem and the necessity for dry periods highlighted the difference between this type of farming and that of the normal farmer in that the farmer irrigating his crops applies water only when necessary. In fact, in a country where water is in short supply, his main aim is to conserve water. Whereas, at a sewage works, the effluent has to be disposed of 365 days of the year with the volume being highest during the rainy season due to infiltration and in actual fact this is the time of the year when the least amount of water is required for agricultural purposes.

Government regulations

At about the same time as these studies were being carried out by the Salisbury City Engineer's Department, the Government showed its concern about what had happened in Lake McIlwaine and other impoundments in the country, and in order to protect the existing and future water resources of Zimbabwe promulgated two sets of regulations to control pollution. The Water Pollution Control (Waste and Effluent Water Standards) Regulations, 1971 (subsequently replaced by the Water (Effluent and Waste Water Standards) Regulations, 1977) dealt with the standards of effluents that may be discharged into natural water courses (see addendum, D. B. Rowe, this volume). In addition the Public Health (Effluent) Regulations, 1972 dealt with the standards required for the re-use of effluents by irrigation. The main features of these Regulations are given in full in the addendum.

As has been seen, the Water (Effluent and Waste Water Standards) Regulations, 1977 are amongst the severest that exist in the world today (see D. B. Rowe, this volume). This meant that some form of tertiary physico-chemical treatment of even the highest quality effluents from conventional sewage treatment plants would be required to satisfy these Regulations. Because of this the initial goal of most local authorities and industries was to increase the size of their existing treatment facilities so that an effluent satisfying the Public Health (Effluent) Regulations, 1972 for irrigation purposes would be produced.

Up until this time the research being carried out by the City Engineer's Department was aimed at reducing the amount of nutrients entering Lake McIlwaine in the hope that this could be done economically and improve the quality of the water in the lake. The advent of these Regulations changed the picture completely in that it now became necessary to irrigate all the effluent arising in the City of Salisbury at all times of the year so that the Regulations would be complied with.

Present farming operations

Because of the above Regulations the two major sewage works in Salisbury were developed so that they were capable of dealing with the entire waterborne sewage of the Salisbury area. Most of the older works were eliminated in order to bring the raw sewage to the two main works. As previous investigations had shown that irrigation was by far the most economical form of tertiary sewage treatment to satisfy the Water (Effluent and Waste Water Standards) Regulations, 1977, a further four farms were purchased by the City of Salisbury for this purpose. The total area of the farming land was now 3411 ha, of which 1397 ha were irrigable and arable. The main crops grown were maize during the summer months and wheat during the winter. The problems of the rainy season and of crop farming continued to persist, however, and so further investigations were made using large areas for pasture and forestry.

These studies revealed that pasture irrigation was by far the most successful method of using sewage effluents to remove nutrients efficiently as the problems of necessary dry periods found in crop irrigation did not exist and that even during the rainy season at least 75% of the nutrients were removed by the surface flow of effluents through the grass. It was then decided to concentrate on pasture irrigation, with cattle running on the farms to act as 'mowing machines' to keep the grass under control and to reduce the costs of pollution control by selling the cattle. Nearly all the irrigable lands were planted with star and kikuyu grasses. The star grass provided rapid coverage, but in most pastures was eventually dominated by the slower growing kikuyu grass. Grass production was more prolific on heavy soils than on sandveld. The pastures are flood irrigated about every three weeks. It has been found that no extra fertilisers have to be used on the heavy soils but that the sandveld required extra fertilisation to increase grass growth and discourage weeds.

Basically, three herds of cattle are run on the farms: namely, a breeding herd of 900 females using Beefmaster bulls in conjunction with Beefmaster semen, a breeding herd of 100 females being bred to Hereford bulls and Hereford semen, and a breeding herd of 600 which is predominantly of the Sussex breed being bred to Sussex bulls and Sussex semen. At the moment, a total of 5376 head of cattle are run on the farms and although the carrying capacity of the pastures has not reached its full potential it is hoped that 7000 head of cattle will be run when the pastures are fully developed.

The lush pastures would seem to be a perfect environment for cattle, but it has been discovered that these pastures are an ideal incubator for all sorts of pests, diseases, etc., and expensive control measures are essential to keep the cattle in good health. Oral dosing is carried out regularly for Liver Fluke and intestinal parasites, with vaccinations for Rift Valley Fever, Lumpy Skin, Vibriosis, Contagious Abortion and Enterotoximia. Two recent severe problems encountered have been Bloat, especially on pastures grown on the heavier soils, and an infestation of snails causing Liver Fluke.

The disposal of digested sludge from the sewage works has been a recognised problem all over the world. Because of fears of Tapeworm (*Taenia saginata*) eggs surviving the mesophilic digestion processes, this sludge was never used on the pasture lands. However, recent laboratory investigations of the viability of *Ascaris* and *Taenia saginata* ova in the digested sludge revealed that none of the ova recovered were viable. As a result of these investigations, certain areas are now being irrigated with digested sludge in admixture with final effluent, cattle are being grazed on this land, and it is hoped that the incidence of 'measly beef' in these cattle will be no higher than that found in cattle which have grazed on pastures irrigated with final effluent only.

In terms of forestry, 73 ha of trees have been planted to investigate their suitability as an irrigated crop. Eight hectares are poplars whilst the rest are blue gums. The trees have grown well on good soils where they have not been irrigated during the summer, but they have been stunted where they have been summer irrigated. Two species of blue gums, *Eucalyptus grandis* and *Eucalyptus camaldulensis*, were grown and the latter was found to be the more successful. As a result, the lands bordering the irrigated pastures and are being planted with *Eucalyptus camaldulensis* to act as a safety barrier between the pastures and nearby water courses. It is interesting to note that the majority of the indigenous trees have not been able to stand up to the intensive irrigation and are slowly dying off.

Siting of the intake tower

The problems arising from the presence of algae in Lake McIlwaine were greatly aggravated by the position of the intake tower and the fact that only three fixed intake levels were available. The tower, which was situated for general convenience and access about 25 m from the dam wall (see N. A. Burke and J. A. Thornton, this volume), lay within a bay formed by the adjacent hills and the dam wall. The prevailing easterly winds blew algae-laden surface water into this bay causing at times very heavy concentrations of algae to accumulate there. This water movement in turn caused a downward current on the upstream face of the dam wall taking active algae in a reverse current along the lower levels of the epilimnion. This process is intermittent, depending on the winds, and the changing concentrations of algae produced corresponding variations in the pH of the water which was normally drawn off at the 7.5 m level. The frequent variation of pH, which could occur extremely rapidly and without warning, required constant surveillance if the correct alum dose was to be maintained at all times and this made the operation of the works more difficult. The effects which arose from the original position of the intake tower were sufficiently serious that consideration was given to the provision of a new point of intake which would be less subject to the influences of wind action on the lake and which would produce a better quality raw water.

An intensive sampling programme was carried out on the lake and a site was chosen for the construction of a new intake tower. This tower was built with four intake levels all of which were at varying heights above the thermocline and one was fitted with a telescopic arrangement which could be raised or lowered by a depth of 2 m.

Since the construction of this tower weekly depth sampling has been carried out at the tower and the periscopic inlet is raised or lowered according to where the best quality raw water is found. The judicial use of this intake has proved to be of enormous benefit to the operation of the water-works and the constant supply of good quality raw water to the works has resulted in a large saving in the cost of chemicals and very good quality treated water being supplied to the consumer.

Recovery of Lake McIlwaine

Other papers in this volume have detailed the physical, chemical and biological changes that have resulted in the lake following the implementation of the nutrient diversion programme by the City of Salisbury. All of these factors confirm that the lake is in the process of recovery, and this is borne out by the fact that treatment of the raw water from Lake McIlwaine has become easier in recent years, with the alum dose being used dropping from an average of 55.0 mg l^{-1} in 1978 to a present-day (1981) average of 30.0 mg l^{-1}. No problems of algae, iron, manganese, tastes or odours are being experienced. As a result, a very high quality treated water is being pumped to the consumers in the City of Salisbury.

Addendum

Prescribed standards of effluents from sewage treatment works for use in crop irrigation as given in the schedule (Section 4) of the Public Health (Effluent) Regulations, 1972, promulgated in Rhodesia Government Notice No. 638 of 1972.

Effluent from sewage treatment works Type of usage or crops	Minimum standards of purity of effluent	Method of irrigation	Other requirements
A. (a) Grain crops; and (b) Crops grown for industrial processing such as oil-seeds, fibre etc. which are not for direct human consumption, but excluding crops grown for dehydration, canning or preserving; and (c) Crops grown solely for seed-production for sale to registered seed merchants but not for human consumption; and (d) Nursery production, excluding cut flowers grown for sale; and (e) Fodder crops for harvesting; and	(1) Biochemical oxygen demand not exceeding 70 parts per million; and (2) Stability as measured by the methylene blue test not less than 36 hours	Surface only	
(f) Pastures for slaughter stock;			No grazing to be permitted within 24 hours of application of effluent, and drinking troughs of potable water to be provided for stock
(g) Deciduous and citrus orchards, trellised vines, plantation and tree crops			No fruit windfalls to be marketed
B. As in A.(a), (b), (c), (d), (e) and (f)	(1) Biochemical oxygen demand not exceeding 30 parts per million; and		

	(2) Stability as measured by the methylene blue test not less than 10 days	Surface or sprinkler	As for A.
C. (a) As in A.; and (b) Pastures for dairy stock; and (c) Cut flowers grown for sale	(1) Biochemical oxygen demand not exceeding 10 parts per million; and		As for A.(f) and (g)
	(2) Stability as measured by the methylene blue test not less than 21 days; and	Surface or sprinkler	
	(3) *E. coli* (type 1) not exceeding 1000 per 100 millilitres		
D. Public amenities, e.g. sports fields, public parks, golf courses, etc. but not swimming pool surrounds	(1) Biochemical oxygen demand not exceeding 10 parts per million; and		
	(2) Stability as measured by the methylene blue test not less than 21 days; and	Surface or sprinkler	
	(3) *E. coli* (type 1) not exceeding 1000 per 100 millilitres; and		
	(4) Residual chlorine not less than 0.3 parts per million after 30 minutes' contact in samples taken at the sewage treatment works		

Acknowledgements

The author thanks Mr L. Mitchell, Director of Works, and the Salisbury City Council for permission to present this paper.

References

Marshall, B. E. and A. C. Falconer, 1973a. Physico-chemical aspects of Lake McIlwaine (Rhodesia), a eutrophic tropical impoundment. Hydrobiol., 42: 45–62.

Marshall, B. E. and A. C. Falconer, 1973b. Eutrophication of a tropical African impoundment (Lake McIlwaine, Rhodesia). Hydrobiol., 43: 109–124.

McKendrick, J., 1973. The physico-chemical treatment of raw sewage. M.Phil. Thesis, University of Rhodesia.

McKendrick, J., 1974. Water pollution and its control in Salisbury. Proc. Rhod. Sci. Congress, Salisbury.

McKendrick, J., 1981. Water supply and wastewater treatment. Zimbabwe Sci. News, 15: 89–91.

McKendrick, J. and R. K. Williams, 1968. The effects of urban drainage on Lake McIlwaine, Rhodesia and subsequent water purification difficulties. Proc. I.W.P.C. (Sth. Afr. Br.) Conf., East London, Republic of South Africa.

Neely, A. B., 1975. Nutrient removal in sewage treatment. M.Phil. Thesis, University of Rhodesia.

Fisheries
K. L. Cochrane

The Parks and Wild Life Act, No. 14 of 1975, provides legislation for, *inter alia*, the preservation, conservation, propagation or control of the fish of Zimbabwe. Part XII, Sections 71 through 83, deals specifically with fish.

Zimbabwe has no natural lakes and relies for its water supply on a large number of man-made lakes. The indigenous fish are therefore riverine in origin although the majority have become well established in dams. The sharp-toothed catfish, *Clarias gariepinus*, is one species which appears not to favour a lacustrine environment and generally shows a decline in numerical importance in large man-made lakes (Marshall, 1977; B. E. Marshall, this volume). Similarly, the indigenous human population of Zimbabwe are not traditional fishermen although the Batonka people of central and north-western Zimbabwe were known to utilise the fish of the Zambesi River to a limited extent (Kenmuir, 1978) and now provide most of the manpower for the inshore fishery of Lake Kariba.

Despite this historical background, the value of the fish resources of Zimbabwe has long been appreciated (Bowmaker, 1975), and three Government bodies (i.e., the Ministries of Agriculture and Home Affairs and the Department of National Parks and Wild Life Management) have been actively involved in the development and promotion of the fish resource. However, there can be little doubt that the fish in Zimbabwe are grossly under-utilised (Bowmaker, 1975).

A large percentage of the area covered by dams in Zimbabwe, as well as much of their shoreline, comes under the control of the Department of National Parks and Wild Life Management and consequently the role of developer of the resource has rested largely with this Department. This has resulted in the fisheries authorities of the Department of National Parks and Wild Life Management having the dual and in some respects contradictory roles of conserving the indigenous fish fauna of Zimbabwe while attempting to promote optimal utilisation of the waters under their control for fish production. This dual role is reflected by the relevant sections of the Parks and Wild Life Act, 1975.

The Parks and Wild Life Act of 1975

Section 79 (1) allows for the control of commercial fishing in Zimbabwe by prohibiting the catching of fish for commercial purposes in any waters of Zimbabwe without the relevant permit and implementation of this law is facilitated by Section 80 (1) which makes the possession of a fishing net by an unauthorised person illegal. Section 83 sets out the legislation for the granting of permits for commercial fishing and, while all commercial fishing in Zimbabwe currently uses nets, allowances are made for the use of other methods such as electro-fishing if the circumstances warrant it.

Recreational fishing is covered by Section 82 of the Act and is limited to conventional rod and line techniques, spear-gun, spear or basket traps throughout the country. Because of the poor visibility in the waters of Lake McIlwaine, line fishing is the only recreational fishing method used regularly on that lake.

The above sections of the Parks and Wild Life Act give the authorities a measure of control over the methods of fish exploitation and Section 73 allows regulation of the level of exploitation. This section empowers the Minister of Natural Resources and Water Development to declare any water body to be 'controlled fishing waters' and subsequently to regulate or prohibit fishing in such waters.

Protection of the indigenous species against disruption by the introduction of exotic species to inland waters is covered by Section 77 of the Act which prohibits such introductions unless undertaken by the relevant authorities. This law is difficult to enforce and illegal introductions, either accidental or by enthusiastic but misguided anglers are not uncommon. *Cyprinus carpio* and *Micropterus salmoides* have both been accidentally introduced into Lake McIlwaine but fortunately neither species has become established and B. E. Marshall (this volume) doubts that either species has bred in the lake.

Implementation of the Act on Lake McIlwaine

There is one licensed commercial fishery operation in Lake McIlwaine which commenced operations in 1956. Surface-set gill nets consisting of 25 m mounted length panels of 76 mm, 102 mm and 127 mm stretched mesh are used. A total of 2000 m to 2500 m of net is set per night (Marshall, 1978). Composition of the commercial catch is described by B. E. Marshall (this volume). The mean total catch per annum from 1972 to 1976 was 110.5 tonnes.

A permit is necessary to fish recreationally on Lake McIlwaine and a charge of approximately 20c per day is made for this permit. Permits are issued by Rangers of the Department of National Parks and Wild Life Management who patrol the shores of the lake. Fishing is permitted throughout the year. Anglers on Lake McIlwaine can be divided into recreational and subsistence fishermen. Both groups make use of rod and line and the major difference between the two groups lies in selectivity; the subsistence angler being content with smaller fish and thus exploiting a wider range of species (Marshall, 1978). Angling pressure has been found to be seasonal with a decrease in effort during the winter months. The maximum number of anglers recorded in a day was 1226 on a week-end in January, with a minimum of 121 on a week-day in June (Marshall, 1978). Numbers were based on visual counts and therefore probably underestimate the actual numbers. The total annual catch by anglers was estimated at 114 tonnes and thus angling has a similar impact on the fish as the commercial fishing. If the total marketing value of the recreational fishery was studied it is likely that this would be of the greatest economic importance and therefore should not be neglected in favour of greater commercial utilisation of the fish.

A third source of fishing pressure is illegal netting, predominantly making use of beach seine nets. Attempts are made to control this and in 1977 alone over 100 arrests were made for illegal netting, but this was considered to be but a small fraction of the illegal fishermen operating (Marshall, 1978). The annual catch taken by these fishermen was estimated to be between 50 and 75 tonnes and therefore may represent between 25 and 35% of the total fish biomass removed from the lake. The nets used may also be having a detrimental effect on the juvenile cichlid population (Marshall, 1978). The illegal fishery therefore represents a serious obstacle in effective management of fish in the lake.

Production potential of the dam

The total yield of fish from Lake McIlwaine was estimated at 120 kg ha^{-1} yr^{-1}

(Marshall, 1978). This yield compares favourably with other freshwater bodies (Table 9). Actual fish production figures are not available for Lake McIlwaine, but phytoplankton productivity has been calculated to be 1.43 kg C m^{-2} yr^{-1} (Robarts, 1979; this volume). Using the model of Oglesby (1977) this should result in a fish production of 20.4 kg ha^{-1} yr^{-1} while the morpheodaphic factor of Henderson and Welcomme (1974) predicts 46 kg ha^{-1} yr^{-1}. While clearly not directly applicable to Lake McIlwaine, these do indicate that exploitation of fish in the lake may be approaching the maximum, and hence the fishing industry on the lake must continue to be carefully monitored. The recreational fishery is likely to be relatively stable and manageable but the presence of a significant illegal fishery makes accurate management difficult. Clearly the ideal would be to attempt to incorporate those presently involved in illegal fishing into the commercial activity thereby facilitating management of the resource and reducing the expenditure on law enforcement.

Table 9 Production of fish from various large water bodies expressed as kg ha^{-1} yr^{-1} (after Bowmaker, 75)

Lake	Production
Lake Tempe, Indonesia	800
Lake McIlwaine, Zimbabwe	120
Lake Quarum, Egypt	104
Lake Malaŵi (eastern arm), Central Africa	50
Lake Tiberias, Israel	40
Lake Victoria, East Africa	2

References

Bowmaker, A. P., 1975. Protein production from fresh water with particular reference to Rhodesia. Rhod. Sci. News, 9: 212–216.

Henderson, H. F. and R. L. Welcomme, 1974. The relationship of yield to morpheodaphic index and numbers of fishermen in African inland fisheries. UN/FAO CIFA Occ. Pap. No. 1.

Kenmuir, D. K., 1978. A wilderness called Kariba. Mardon, Salisbury.

Marshall, B. E., 1977. On the status of *Clarias gariepinus* (Burchell) in large man-made lakes in Rhodesia. J. Limnol. Soc. sth. Afr., 3: 67–68.

Marshall, B. E., 1978. An assessment of fish production in an African man-made lake (Lake McIlwaine, Rhodesia). Trans. Rhod. Scient. Ass., 59: 12–21.

Oglesby, R. T., 1977. Relationships of fish yield to lake phytoplankton standing crop, production and morphoedaphic factors. J. Fish. Res. Bd. Can., 34: 2271–2279.

Robarts, R. D., 1979. Underwater light penetration, chlorophyll *a* and primary production in a tropical African lake (Lake McIlwaine, Rhodesia). Arch. Hydrobiol., 86: 423–444.

Recreation
G. F. T. Child and J. A. Thornton

Lake McIlwaine is one of the most popular recreational venues in Zimbabwe (Child, 1977). The close proximity of the Robert McIlwaine National Recreational Park to the City of Salisbury makes the lake an ideal site for day visitors from the capital as well as for tourists and other visitors contemplating a more extended stay. A wide variety of interests are catered for within the Park, although most of these are aquatically oriented, and a choice of Government-run and commercially-operated facilities is to be had. A number of clubs and non-profit-making organisations also have facilities within the National Park to serve their members. It can easily be said that Lake McIlwaine and the Robert McIlwaine Recreational Park have something for everyone.

Van Hoffen *et al.* (1979) have undertaken a survey of the recreational potential of the lake and its environs, and have distinguished five major categories of recreational usage (Table 10). Foremost amongst these categories is that of angling. Boating (including sailing, power-boating and rowing) is the second most extensively catered for type of recreation, followed by general aquatic recreation such as swimming, and by game viewing. Included in Table 10 are the service industries (those offering accommodation) which form a major part of the amenities of the Park.

Because of the fact that the game park on the south bank of the lake is of necessity fenced, and due to the fact that all visitors entering the game park pay an entrance fee, the user statistics for that portion of the Recreational Park are well documented. These statistics thus form the bulk of this section.

Table 10 Recreational sites in the Robert McIlwaine National Recreational Park at Lake McIlwaine (after Van Hoffen *et al.*, 1979)

Activity	No. of sites
Angling	26
Boating (all forms)	17
sailing/yachting	12
power-boating	16
rowing/canoeing	15
Swimming and other water sports	14
Accommodation (all forms)	19
hotels/lodges, etc.	10 (90 units)
camping/caravans	12 (353 units)
Game viewing	1
Total number of recreational sites	33

To supplement these figures, a limited survey of the major north bank users (both commercial and club facilities) was undertaken during 1980 to determine the approximate recreational usage of that portion of the lake side Park. Thus, whilst this section cannot be said to be comprehensive, it can claim to give a true picture of the recreational utilisation of the Robert McIlwaine National Park. The recreational categories of Van Hoffen *et al.* (1979) have been used in the following discussion.

Angling

Marshall (1978) undertook a survey of angling usage of the lake during 1974–76. Whilst his data do not show any changing patterns of recreational usage they do show that the lake is a popular fishing venue. During the two years of the study nearly 5500 recreational anglers per annum used the facilities provided; the majority of the recreational anglers (nearly 5000) visited the lake at week-ends whilst the remainder fished during the week. The success rate of anglers ranged from 8% during July (mid-winter) to 70% in April (late summer; the breeding period for most species of fish in Lake McIlwaine). The mean success rate was 48% for the year with the average mass of fish per angler being in the region of 1.5 kg. Based on these angling figures, Marshall (1978) calculated the annual production of the lake at 62 metric tonnes.

Recreational anglers at Lake McIlwaine catch imberi (*Alestes imberi*), greenheaded tilapia (*Sarotherodon macrochir*), and redbreasted tilapia (*Tilapia rendalli*) predominantly at a rate of 3 to 4 fish per angler (Marshall, 1978; B. E. Marshall, this volume). The commercial fishermen and subsistence anglers do not compete for these species. Thus, recreational angling in Lake McIlwaine assists in the management of the lake fishery through the cropping of some of the common species not taken by other fishing methods, in addition to providing recreation for large numbers of fishermen.

Boating

Lake McIlwaine is a popular boating venue with numerous facilities for launching, hiring and storing water-borne craft of all descriptions. Most lake-side facilities cater for all types of vessels, although a few specialise in either sailing or power-boating.

The Mazoe Yacht Club and the Jacana Club are the largest sailing facilities.

These clubs had a combined membership of approximately 750 in 1980, who owned a total of 250 yachts. Both clubs organise annual regattas and courses for junior sailors.

Power-boating and rowing are largely commercial activities on the lake, with the Ancient Mariner and Admiral's Cabin Boat Stations accounting for the bulk of the business. Launching and storage of privately owned boats and hire of motor-boats, canoes and row-boats are undertaken by both firms, together with some limited repair and maintenance work. Some 300 boats were stored at each of these sites and 90 boats were available for hire at each of these sites at the time of writing.

Peak periods of lake usage by boat traffic are at the week-ends, and more especially during the extended holiday week-ends of Christmas and New Year, Easter, and the mid-year Hero's (formerly Rhodes' and Founders') Days (Van Hoffen *et al.*, 1979). The annual school holidays (December/January, April/May, and August/September) are also popular times for boating. Numbers of vessels afloat during these periods can range up to several hundred craft. Mid-winter is the least popular time for boating (Van Hoffen *et al.*, 1979).

Swimming

Most of the larger club sites and commercial facilities on the lake-shore provide swimming pools and children's playgrounds for their patrons. Pools are also provided at the National Park rest camp. Swimming also takes place in the lake itself with a number of sandy beaches being sprayed periodically with a molluscicide to discourage the spread of schistosomiasis (Bilharziasis). Water skiing is a popular sport which can be combined with power-boating, and a number of aquatic slalom competitions have been held. Other popular water sports include SCUBA diving and wind-surfing, both of which have become more popular in recent years.

Accommodation

There are nearly 450 units of accommodation around Lake McIlwaine; the majority of the accommodation being in the form of permanently-sited caravans within the areas operated by the various clubs (Van Hoffen *et al.*, 1979). Accommodation available to the public is largely distributed between the Hunyani Hills Hotel (L**) on the north bank and the Government-run chalets and lodges within the game park on the south bank. Numerous sites for camping and caravanning are also provided.

Figure 1 shows the changes in usage of the camping and caravanning facilities and permanent, furnished accommodation provided by the Department of National Parks and Wild Life Management (Child, 1980), for the period between 1960 and 1980 at Lake McIlwaine. Whilst the data shown in Fig. 1 may not be a true reflection of the overall usage of the Recreational Park, it does show certain trends in usage which may apply to the commercial facilities as well. Certainly the north bank commercial facilities are used by much higher numbers of people than the south bank facilities (Zimbabwe National Tourist Board, personal communication).

Usage of both camping and permanent accommodation in the National Park reached a fairly constant level of approximately 18,400 people per annum during the 1960s, with the permanent accommodation catering for a slightly higher overall percentage of the trade than the camping sites, although the relative popularity of the two types of accommodation varied from year to year. Between 1968 and 1969, usage of the accommodation facilities in the Park plummetted to an average of 3600 people per annum, with the permanent accommodation retaining a slightly higher proportion of the trade.

The reasons for this drop in utilisation might at first glance appear to be related to the increasing eutrophication of the lake; the full effects of which became apparent at about the time of the decline in usage (Marshall and Falconer, 1973; N. A. Burke and J. A. Thornton, this volume). This is however unlikely in light of other factors which affected the market: namely,

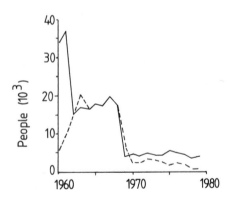

Fig. 1 Thousands of people using the camping (---) and permanent accommodation (—) operated by the Department of National Parks and Wild Life Management at the Robert McIlwaine National Park.

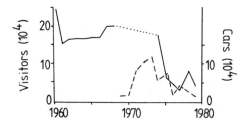

Fig.2 Tens of thousands of persons visiting the National Park as day visitors to the south bank only (——); and the number of automobiles recorded at the game park (– – –). Dotted line (. . . .) indicates data were not available.

the introduction of petrol rationing in 1968. Rationing was lifted briefly between 1971 and 1974, and remained in force until 1980 (Ministry of Commerce and Industry, personal communication). Combined with this, a worsening security situation resulted in a redcution of available recreational time, particularly after the introduction of general conscription in 1974. Thus, it would seem most likely that an artificial alteration of the public's attitude toward recreation brought about the decline in usage of the accommodation within the Park, rather than the effects of eutrophication, although the latter must have been initially contributory.

Game-viewing

The above conclusions are largely supported by the data shown in Fig. 2 (Child, 1980), which shows the usage of the game park by day visitors together with the automobile traffic through the Park between 1960 and 1980. The vehicular traffic in particular shows the effects of petrol rationing quite clearly. Fewer cars used the Park during the periods of rationing than during the brief period when rationing was lifted (1971–74). Fig. 2 also suggests that the number of day visitors to the Park only decreased after the introduction of conscription (1974), as a result of the reduced time available for recreation; confirmatory data from the period 1969 to 1973 are unfortunately not available. It must, however, again be stressed that these data are for the areas of the Recreational Park directly controlled by the Department of National Parks and Wild Life Management and may not reflect the utilisation of the Park as a whole.

A comparison of Figs. 1 and 2 shows that most people prefer to visit the

Park for relatively short periods. On average 182,000 people visited the Recreational Park annually between 1960 and 1975 compared to 25,000 who utilised the accommodation facilities during the same period. Afer 1975, similar proportions of people visiting the Park were recorded albeit at reduced total numbers. Generally, day visitors travel to the Park to enjoy a day's game-viewing and have a picnic lunch at one of the several picnic sites provided. Fishing and bird-watching are also popular pastimes for day visitors, in addition to game-viewing.

Conclusions

It is readily apparent from the foregoing that Lake McIlwaine and the Robert McIlwaine National Recreational Park are heavily utilised recreational venues. Several hundred thousand people, a fairly high percentage of the total population of the City of Salisbury, use the Park and its facilities annually (Zimbabwe National Tourist Board, personal communication), despite a major decline in recreational usage in recent years. This decline has been shown to be due in large part to artificial alterations of the public's recreational habits by petrol rationing and war. Now both of these deleterious influences have been lifted, it is to be anticipated that usage of the Park and its facilities will again increase, particularly in light of the very substantial capital investment in equipment and facilities that has been made by major lake users over the years. It is encouraging to note that such an up-swing in recreational usage is already being noticed at the time of writing (Salisbury Publicity Association, personal communication).

References

Child, G. F. T., 1977. Annual report of the Director of National Parks and Wild Life Management for 1977. Department of National Parks and Wild Life Management, Salisbury.
Child, G. F. T., 1980. Annual report of the Director of National Parks and Wild Life Management for 1980. Department of National Parks and Wild Life Management, Salisbury.
Marshall, B. E., 1978. An assessment of fish production in an African man-made lake (Lake McIlwaine, Rhodesia). Trans. Rhod. Scient. Ass., 59: 12–21.
Marshall, B. E. and A. C. Falconer, 1973. Eutrophication of a tropical African impoundment (Lake McIlwaine, Rhodesia). Hydrobiol., 43: 109–124.
Van Hoffen, P., F. R. Chinembiri, S. R. T. Manyande, K. N. Mutero, J. A. Njunga, R Patel, J. H. Sobantu and E. T. Theunissen, 1979. Water-based recreation in the Hunyani-Makabusi basin. M. A. Thesis, Carnegie-Mellon University.

Research: perspectives
J. A. Thornton

The research effort on Lake McIlwaine has in the past been motivated by two key questions. The first of these, which dominated what may be called the first phase of the research programme on Lake McIlwaine, was what is the cause of the eutrophication of the lake? The effects of the eutrophication process were to some extent obvious and have been outlined in several of the preceeding papers. The second phase of the Lake McIlwaine research programme was aimed at answering the question what can be done to retard or reverse the eutrophication process? Implicit in this question was the query what effect will eutrophication control measures have on the lake? These questions too have been answered in the preceeding papers. Thus, the research effort on Lake McIlwaine stands at a cross-roads, and it remains for this section to assess the foregoing reviews and to examine the research needs of the future. And indeed it was this need for an assessment of the research programme on Lake McIlwaine that provided the impetus for this volume.

Whilst the trophic relationships between the communities of organisms described in this monograph are largely beyond the scope of this volume, by way of a summary in this section some mention will be made of the lake ecosystem as a whole and the interactions between its components. Both the eutrophication and recovery of Lake McIlwaine will be outlined in order to highlight those areas where further research effort is required and those areas where the research effort has been sufficient to meet the needs. Reference in this section will be to the relevant review papers included in this volume which in turn are based on the original research papers.

The eutrophication of Lake McIlwaine

When it was created, Lake McIlwaine was a mesotrophic impoundment similar in many respects to the existing reservoirs on the Hunyani River (J. McKendrick, this volume). Nitrogen and phosphorus concentrations were relatively low (J. A. Thornton and W. K. Nduku, this volume) and algal growth was probably phosphorus limited (C. J. Watts, this volume). The phytoplankton community was dominated by cyanophytes, particularly *Microcystis aeruginosa* Kutz, but species diversity was great and a large number of chlorophytes were present (J. A. Thornton, this volume). The zooplankton of the lake was dominated by the cladoceran *Ceriodaphnia dubia* Richard (J. A. Thornton and H. J. Taussig this volume). Zooplankton formed the major food source of the dominant fish species, *Clarias gariepinus* (Burchell), and the fish fauna of the lake had the predominantly riverine characteristics of a newly created man-made lake (B. E. Marshall, this volume). The oligochaete *Branchiura sowerbyi* Beddard formed the major component of the benthic fauna (B. E. Marshall, this volume) and large amounts of aquatic macrophytes were present on and around the lake (M. J. F. Jarvis *et al.*, this volume).

Throughout this formative period, rainfall was good and lake levels were reasonably constant (B. R. Ballinger and J. A. Thornton, this volume) although drought years did occur in 1960 and 1964 when lake levels dropped by several meters. During this period also (1960–64) the volume of municipal wastewater being treated by the Salisbury sewage works rose considerably (J. McKendrick, this volume; K. Munzwa, this volume) and the first manifestations of eutrophication were being observed in the lake (N. A. Burke and J. A. Thornton, this volume). The process of eutrophication reached its peak by 1968 when a further period of low rainfall occurred; the effects of eutrophication were probably heightened by the increased sewage flows and low rainfall.

By 1968 nitrogen and phosphorus concentrations in the lake had increased by a factor of between 5 and 10 times their original levels (J. A. Thornton and W. K. Nduku, this volume), and nitrogen became the primary growth-limiting nutrient (C. J. Watts, this volume). The phytoplankton of the lake was dominated by *Microcystis aeruginosa* and species diversity was low. *Microcystis, Anabaena flos-aquae* (Lyng.) Breb. and *Melosira granulata* (Ehr.) almost exclusively made up the planktonic flora (J. A. Thornton, this volume). Significantly, no nitrogen fixation was detected in the impoundment (R. D. Robarts, this volume). The fish fauna had taken on more lacustrine characteristics by this time, and the largely phytoplanktivorous *Tilapia rendalli* (Boulenger) and *Sarotherodon macrochir* (Boulenger) now dominated (B. E. Marshall, this volume). Chironomids, especially *Chironomus* sp., formed the largest component of the benthic fauna (B. E. Marshall, this volume) and an apparent shift in zooplankton dominance also took place, with calanoid copepods becoming more prevalent (J. A. Thornton and H. J. Taussig, this volume). Following the application in 1969 of the herbicide 2,4-D to control the spread of *Eichhornia crassipes* (Mart.) Solms, the macrophyte flora of the lake was substantially altered (M. J. F. Jarvis *et al.*, this volume).

The massive algal blooms, characteristic of eutrophic lakes, together with the accumulation of inorganic ions in the lake hypolimnion, created water treatment problems which led to difficulty in meeting the demand for potable water (J. McKendrick, this volume). Because of the fact that initial research programmes pointed to the excessive nutrient input, largely derived from treated municipal wastewater released by the City of Salisbury, as a significant factor in the eutrophication of Lake McIlwaine, the City began investigating ways of reducing the amount of phosphorus and nitrogen in the sewage effluent (J. McKendrick, this volume). Meanwhile in 1970 Government promulgated stringent water pollution control legislation (D. B. Rowe, this volume; J. McKendrick, this volume) which legally required the City to improve the quality of its effluent.

The recovery of Lake McIlwaine

The second phase of the research programme on Lake McIlwaine thus began in 1970 with the phased implementation of a nutrient diversion programme by the City of Salisbury (J. McKendrick, this volume). Because of the phased nature of the nutrient diversion programme (which took place between 1970 and 1975), little change was noted in the trophic status of the lake until 1976 (R. D. Robarts, this volume; J. McKendrick, this volume; J. A. Thornton and W. K. Nduku, this volume). During the years 1976–78, phosphorus concentrations in the lake decreased to pre-eutrophic levels whilst nitrogen concentrations remained largely unchanged (J. A. Thornton and W. K. Nduku, this volume) despite a large increase in the volume of wastewater, and hence in the nutrient loads, entering the sewage works for treatment (J. McKendrick, this volume). Because of the use of pasture irrigation as a tertiary sewage treatment method (J. McKendrick, this volume) the nitrogen fractions entering the lake now consisted of more nitrate than ammonia (J. A. Thornton and W. K. Nduku, this volume); this in itself may have led to a reduction in the intensity of the algal blooms experienced in the lake as nitrate may be toxic to *Microcystis aeruginosa* (Kappers, 1980). In addition, phytoplankton growth became limited by phosphorus during this period (C. J. Watts, this volume). Despite the decline in the intensity of algal blooms and the shift in algal growth-limiting nutrient, no significant decrease in primary production has been recorded (R. D. Robarts, this volume), largely due to the improved light climate (R. D. Robarts, this volume; B. E. Marshall, this volume). There has been a limited re-establishment of the macrophyte flora of the lake margins (M. J. F. Jarvis *et al.*, this volume) which, combined with the improved light climate, led to the increased growth of benthic and epiphytic algae (B. E. Marshall, this volume). This may have been contributory to the increased importance of *Labeo altivelis* Peters within the fish fauna as this fish is phytoplanktivorous and feeds largely on periphyton B. E. Marshall, this volume). No major changes have yet been noted in the phytoplankton of the lake (J. A. Thornton, this volume) but there are indications that the zooplankton are now dominated by cyclopoid copepods (J. A. Thornton and H. J. Taussig, this volume). Nothing is known of any changes which may have taken place in the benthic fauna of the lake (B. E. Marshall, this volume).

That the water quality of the lake has improved is shown by the reduction in the chemical requirements of the water treatment works and the easing of the water supply problems (J. McKendrick, this volume). Present indications are that the lake has returned to a trophic status similar to that prior to the onset of severe eutrophication, and the impoundment has since been referred to as mesotrophic (J. A. Thornton and W. K. Nduku, this volume). As the recovery of the lake has taken place during a period of above average rainfall, some question has arisen as to the effect of a poor rainfall period on the trophic status of the lake, but data obtained during the 1979–80 season indicate that this mesotrophic state will nevertheless be maintained (J. A. Thornton and W. K. Nduku, this volume; Thornton, 1980a). However, the mesotrophic state could be threatened by the increasing inputs of nutrients from diffuse sources (mainly urban stormwater run-off) within the lake catchment (J. A. Thornton and W. K. Nduku, this volume).

Future research

This then is the situation at the time of writing. There are obviously several areas where further research is required; such as, the physics of the lake (P. R. B. Ward, this volume), zooplankton and secondary production (J. A. Thornton and H. J. Taussig, this volume), and the benthic fauna since nutrient diversion (B. E. Marshall, this volume). Further work on pesticides and heavy metals (Y. A. Greichus, this volume; R. S. Hatherly and K. A. Viewing, this volume), on the phytoplankton and primary production (J. A. Thornton, this volume; R. D. Robarts, this volume), and on the macrophytes of the lake (M. J. F. Jarvis *et al.*, this volume) may also be desirable. Aspects of the lake fisheries (B. E. Marshall, this volume; K. L. Cochrane, this volume) are well covered and the chemical water quality is subject to on-going monitoring (J. McKendrick, this volume; D. B. Rowe, this volume). These research needs have been reviewed by Thornton (1980b) and the above-mentioned areas will form the basis of future research programmes.

References

Kappers, F. I., 1980. The cyanobacterium *Microcystis aeruginosa* Kg. and the nitrogen cycle of the hypertrophic Lake Brielle (Netherlands). *In:* J. Barica and L. R. Mur, Hypertrophic ecosystems. Developments in Hydrobiology, 2: 37–43.

Thornton, J. A., 1980a. The Water Act, 1976, and its implications for water pollution control: case studies. Trans. Zimbabwe Scient. Ass., 60: 32–40.

Thornton, J. A., 1980b. A review of limnology in Zimbabwe: 1959–1979. NWQS Rep. No. 1, Ministry of Water Development and Department of National Parks and Wild Life Management, Causeway, Zimbabwe.

7 Bibliography

Margaret J. Thornton and J. A. Thornton

This annotated bibliography spans the years between 1960 and 1981, but most of the publications date from 1973 onwards and reflect the creation of the Hydrobiology Research Unit at the University of Zimbabwe. The main section of the bibliography lists those works which have appeared in the scientific literature, whilst higher degree theses are given in the addendum. Unpublished Government and University reports, whilst numerous, have not been included as many of these were of limited distribution and hence are not widely available outside of their parent institutions.

Bell-Cross, G., 1976. The fishes of Rhodesia. National Museums and Monuments Pub., Salisbury.
  Keywords: species lists, distribution.
Borrett, R., 1969. Flamingos at Lake McIlwaine. The Honeyguide, 59: 37.
  Keywords: Lake McIlwaine, *Phoeniconaias minor, Phoenicopterus ruber,* sighting report.
Bowmaker, A. P., 1975. Protein production from fresh water, with particular reference to Rhodesia. Rhod. Sci. News, 9 (7): 212–216.
  Keywords: water resources, African lakes, fish production, aquaculture.
Campbell, N. A., 1969. Gull-billed Tern (?) at Lake McIlwaine' The Honeyguide, 59: 30.
  Keywords: Lake McIlwaine, *Gelochelidon nilotica,* sighting report.
Caulton, M. S., 1975. Diurnal movement and temperature selection by juvenile and sub-adult *Tilapia rendalli* Boulenger (Cichlidae). Trans. Rhod. Scient. Ass., 56 (4): 51–56.
  Keywords: Lake McIlwaine, littoral, laboratory, *Tilapia rendalli,* daily movement, juvenile, sub-adult, temperature selection.
Caulton, M. S., 1976. The importance of pre-digestive food preparation to *Tilapia rendalli* Boulenger when feeding on aquatic macrophytes. Trans. Rhod. Scient. Ass., 57 (3): 22–28.
  Keywords: *Tilapia rendalli,* laboratory, *Ceratophyllum demersum,* alimentary canal, pharyngeal teeth.
Caulton, M. S., 1977a. A quantitative assessment of the daily ingestion of *Panicum repens* L. by *Tilapia rendalli* Boulenger (Cichlidae) in Lake Kariba. Trans. Rhod. Scient. Ass., 58 (6): 38–42.
  Keywords: Lake Kariba, *Tilapia rendalli, Panicum repens,* feeding.

Caulton, M. S., 1977b. The effect of temperature on routine metabolism in *Tilapia rendalli* Boulenger. J. Fish. Biol., 11: 549–53.
Keywords: *Tilapia rendalli,* respiration, temperature, metabolism.
Caulton, M. S., 1978a. Tissue depletion and energy utilisation during routine metabolims by sub-adult *Tilapia rendalli* Boulenger. J. Fish. Biol., 13: 1–6.
Keywords: *Tilapia rendalli,* ammonia excretion, fat, protein, condition factors, metabolism.
Caulton, M. S., 1978b. The importance of habitat temperatures for growth in the tropical cichlid *Tilapia rendalli* Boulenger. J. Fish. Biol., 13: 99–112.
Keywords: *Tilapia rendalli,* feeding, temperature, *Ceratophyllum demersum,* assimilation efficiency, oxygen consumption, ammonia excretion, energy budget.
Caulton, M. S. and E. Bursell, 1977. The relationship between changes in condition and body composition in young *Tilapia rendalli* Boulenger. J. Fish. Biol., 11: 143–150.
Keywords: *Tilapia rendalli,* condition factors, fat, protein, water, regression analysis.
Clay, D., 1976. Several models for water-air temperature relationships of some African lakes. Water SA, 2 (2): 61–66.
Keywords: Lake McIlwaine, southern Africa, West Africa, water temperature, air temperature, regression analysis.
Clay, D., 1979a. Sexual maturity and fecundity of the African catfish. (*Clarias gariepinius*) with an observation on the spawning behaviour of the Nile catfish (*Clarias lazera*). J. Linn. Soc. Lond., 65: 351–365.
Keywords: *Clarias gariepinus,* fecundity, sexual maturity, breeding, spawning, *Clarias lazera.*
Clay, D., 1979b. Population biology, growth and feeding of African catfish (*Clarias gariepinus*) with special reference to juveniles and their importance in fish culture. Arch. Hydrobiol., 87 (4): 453–482.
Keywords: *Clarias gariepinus,* Lake McIlwaine, temperature, feeding, distribution, population structure, fisheries, spawning, growth, aquaculture.
Clay, D., 1981. Utilization of plant materials by juvenile African Catfish *(Clarias gariepinus)* and its importance in fish culture. J. Limnol. Soc. sth. Afr., 7 (2): 47–56.
Keywords: *Clarias gariepinus,* Lake McIlwaine, feeding, metabolism, digestion, energy budget, respiration, aquaculture.
Evans, P. J., 197. Letter to Editor. The Honeyguide, 69 (97): 37.
Keywords: Lake McIlwaine, *Egretta vinaeceigula,* sighting report.
Falconer, A. C., 1970. Eutrophication and Lake McIlwaine. Rhod. Sci. News, 4: 52.
Keywords: Lake McIlwaine, nutrients, pollution, phytoplankton.
Falconer, A. C. and B. E. Marshall, 1969. Limnological investigations of Lake McIlwaine. Limnol. Soc. Sth. Afr. Newsl., 13 (suppl.): 66–69.
Keywords: Lake McIlwaine, eutrophication, phytoplankton, benthos.
Greichus, Y. A., A. Greichus, H. A. Draayer and B. E. Marshall, 1978. Insecticides, polychlorinated biphenyls and metals in African lake ecosystems. II. Lake McIlwaine, Rhodesia. Bull. Environ. Contam. Toxicol., 19 (4): 444–453.
Keywords: Lake McIlwaine, insecticides, PCB's, heavy metals, organo-chlorine pesticides, water, sediments, plankton, fish, benthos, birds.
Henwood, P., 1973. Openbill Storks at Lake McIlwaine. The Honeyguide, 63 (75): 29.
Keywords: Lake McIlwaine, *Anastomus lamelligerus,* sighting report.
Horne, A. J. In: W. D. P. Stewart, 1974. Blue-green algae. In: A. Quispel, The biology of nitrogen fixation, Elsevier, New York.
Keywords: Lake McIlwaine, nitrogen fixation, ^{15}N, nitrogenase activity.

Jacot Guillarmod, A., 1979. Water weeds in southern Africa. Aquat. Bot., 6: 377–391.
  Keywords: southern Africa, *Eichhornia crassipes, Myriophyllum aquaticum, Salvinia molesta, Azolla filiculoides,* distribution, history, biology, economics, indigenous species, legislation, use.
Jarvis, M. J. F., 1976. Cape Shoveller at Lake McIlwaine. The Honeyguide, 66 (86): 43.
  Keywords: Lake McIlwaine, *Anas smithii,* sighting report.
Jarvis, M. J. F., 1980. Distribution and abundance of waterfowl (Anatidae) in Zimbabwe. Proc. Fifth Pan African Ornithol. Congr., Malawi.
  Keywords: Zimbabwe, Lake McIlwaine, man-made lakes, Anatidae, sighting report, species list, distribution.
Jarvis, M. J. F., M. I. van der Lingen and J. A. Thornton, 1981. Water hyacinth. Zimbabwe Sci. News, 15 (4): 97–99.
  Keywords: Lake McIlwaine, *Eichhornia crassipes,* problems, growth, control.
Jubb, R. A., 1967. Freshwater fishes of southern Africa. Balkema, Cape Town.
  Keywords: species list, distribution.
Junor, F. J. R., 1969. *Tilapia melanopleura* Dum. in artificial lakes and dams in Rhodesia with special reference to its undesirable effects. Rhod. J. agric. Res., 7: 61–69.
  Keywords: *Tilapia melanopleura,* distribution, exotic species, Lake Kyle, Umshandige Dam, Lakes Ngesi, Lake McIlwaine, Lundi River, feeding.
Kenmuir, D. H. S., 1980. Seasonal breeding activity in freshwater mussels (Lamellibranchia: Unionacae) in Lake Kariba and Lake McIlwaine, Zimbabwe. Trans. Zimbabwe Scient. Ass., 60 (4): 18–23.
  Keywords: Lake McIlwaine, Lake Kariba, *Mutela dubia, Caelatura mossambiciensis, Aspatharia wahlbergi,* breeding.
Loewenson, S. W., 1974. Water resources in Rhodesia. Rhod. Sci. News, 8: 354–359.
  Keywords: water resources, Rhodesia, distribution, utilisation.
Loveridge, J. P. and G. Graye, 197. Cocoon formation in two species of southern African frogs. S. A. J. Sci., 75 (1): 18–20.
  Keywords: southern Africa, *Pyxicephalus adspersus, Leptopelis bocogei,* dessication, humidity, Lake McIlwaine, Transvaal, Zululand, cocoons, membranes.
Maar, A. A., 1962. *Marcusenius smithers:* sp. nov. and *Gnathonemus rhodesianus:* sp. nov. (Mormyridae) from the Zambesi River system and *Barbus hondeensis:* sp. nov. (Cyprinidae) from the Pungwe River. Occ. Pap. Nat. Mus. S. Rhod., 3: 780–784.
  Keywords: Lake McIlwaine, *Marcusenius rhodesianus,* species list, distribution, *Marcusenius smithers, Barbus hondeensis,* Pungwe River.
Magadza, C. H. D., 1977a. A note on Entomostraca in samples from three dams in Rhodesia. Arnoldia (Rhod.), 8 (14): 1–4.
  Keywords: Lake McIlwaine, Mazoe Dam, Connemara Dam, *Thermodiaptomus syngenes,* microcrustaceans, zooplankton, species list.
Magadza, C. H. D., 1977b. Determination of development period at various temperatures in a tropical cladoceran; *Moina dubia* DeGuerne and Richard. Trans. Rhod. Scient. Ass., 58 (4): 24–27.
  Keywords: *Moina dubia,* Kafue Gorge Dam, temperature, development times, sewage works.
Magadza, C. H. D. and P. Z. Mukwena, 1979. Determination of the post-embryonic development period in *Thermocyclops neglectus* (Sars) using cohort analysis in batch cultures. Trans. Rhod. Scient. Ass., 59 (6): 41–45.
  Keywords: *Thermocyclops neglectus,* Lake McIlwaine, development times, temperature, laboratory, culture.

Manson, A. J. and C. Manson, 1969. Gull-billed Tern – further sighting. The Honeyguide, 59: 31.
Keywords: Lake McIlwaine, *Gelochelidon nilotica*, sighting report.
Marshall, B. E., 1970. The ecology of the bottom fauna of Lake McIlwaine. Rhod. Sci. News, 4: 53.
Keywords: Lake McIlwaine, benthos, pollution.
Marshall, B. E., 1972. Some effects of organic pollution of benthic fauna. Rhod. Sci. News, 6 (5): 142–145.
Keywords: Makabusi River, Hunyani River, benthos, pollution, biological indicators.
Marshall, B. E., 1974. Notes on the spatial distribution and substrate preferences of *Branchiura sowerbyi* (Oligochaeta: Tubificidae). News Lett. Limnol. Soc. Sth. Afr., 21: 12–15.
Keywords: Lake McIlwaine, distribution, substrate, colour, light.
Marshall, B. E., 1975. Observation on the freshwater mussles [sic] (Lamellibranchia: Unionacae) of Lake McIlwaine, Rhodesia. Arnoldia (Rhod.), 7 (16): 1–16.
Keywords: Lake McIlwaine, species list, Lamellibranchia, distribution, population, growth.
Marshall, B. E., 1977. On the status of *Clarias gariepinus* (Burchell) in large man-made lakes in Rhodesia. J. Limnol. Soc. Sth. Afr., 3 (2): 67–68.
Keywords: Lake McIlwaine, man-made lakes, *Clarias gariepinus*, biology, fisheries.
Marshall, B. E., 1978a. Aspects of the ecology of benthic fauna in Lake McIlwaine, Rhodesia. Freshwat. Biol., 8 (3): 241–249.
Keywords: Lake McIlwaine, benthos, pollution, distribution.
Marshall, B. E., 1978b. Lake McIlwaine after twenty-five years. Rhod. Sci. News, 12 (3): 79–82.
Keywords: Lake McIlwaine, pollution, algae, nutrients, macrophytes, benthos, fish.
Marshall, B. E., 1978c. An assessment of fish production in an African man-made lake (Lake McIlwaine, Rhodesia). Trans. Rhod. Scient. Ass., 59 (3): 12–21.
Keywords: Lake McIlwaine, fisheries, production, angling.
Marshall, B. E., 1979. Observations on the breeding biology of *Sarotherodon macrochir* (Boulenger) in Lake McIlwaine, Rhodesia. J. Fish. Biol., 14 (4): 419–424.
Keywords: Lake McIlwaine, *Sarotherodon macrochir*, breeding, biology, temperature.
Marshall, B. E., 1981. Fish and eutrophication in Lake McIlwaine. Zimbabwe Sci. News, 15 (4): 100–102.
Keywords: Lake McIlwaine, pollution, fish kills, species changes, production.
Marshall, B. E. and A. C. Falconer, 1973a. Physico-chemical aspects of Lake McIlwaine (Rhodesia), a eutrophic tropical impoundment. Hydrobiol., 42 (1): 45–62.
Keywords: Lake McIlwaine, hydrology, temperature, oxygen, nutrients, water quality.
Marshall, B. E. and A. C. Falconer, 1973b. Eutrophication of a tropical African impoundment (Lake McIlwaine, Rhodesia). Hydrobiol., 43 (1/2): 109–123.
Keywords: Lake McIlwaine, pollution, production, water quality, rivers, nutrient budget.
Marshall, B. E. and C. A. Lockett, 1976. Juvenile fish populations in the marginal areas of Lake McIlwaine, Rhodesia. J. Limnol. Soc. Sth. Afr., 2 (2): 37–42.
Keywords: Lake McIlwaine, species composition, production, distribution, littoral zone.
Marshall, B. E. and J. T. van der Heiden, 1977. The biology of *Alestes imberi* Peters (Pisces: Characidae) in Lake McIlwaine, Rhodesia. Zool. Afr., 12 (2): 329–346.
Keywords: Lake McIlwaine, *Alestes imberi*, breeding, feeding, condition factors, growth, biology.
McKendrick, J., 1979. Compulsory re-use of water due to very strict water pollution control regulations in Salisbury, Rhodesia. Proc. Water Re-use Symp., Washington, D. C., 2: 1035–1048A.

Keywords: Lake McIlwaine, water re-use, pollution control, water quality, water treatment, sewage treatment, legislation.

McKendrick, J., 1981. Water supply and wastewater treatment. Zimbabwe Sci. News, 15 (4): 89–91.
Keywords: Lake McIlwaine, wastewater treatment, pollution control, water supply, legislation.

Mitchell, D. S. and B. E. Marshall, 1974. Hydrobiological observations on three Rhodesian reservoirs. Freshwat. Biol., 4: 61–72.
Keywords: Lake McIlwaine, Mazoe Dam, Mwenje Dam, water quality, diurnal variation, phytoplankton, primary production, trophic status.

Munro, J. L., 1966. A limnolgical survey of Lake McIlwaine, Rhodesia. Hydrobiol., 28: 281–308.
Keywords: Lake McIlwaine, topography, hydrology, temperature, water quality, macrophytes, phytoplankton, zooplankton, benthos, fish, trophic status.

Munro, J. L., 1967. The food web of a community of East African freshwater fishes. J. Zool., Lond., 151: 389–415.
Keywords: Lake McIlwaine, *Clarias gariepinus*, *Sarotherodon macrochir*, *Marcusenius macrolepidotus*, *Hydrocynus vittatus*, *Sarotherodon mossambicus*, feeding, benthos, insects, molluscus, fish, zooplankton, algae, detritus.

Nduku, W. K., 1976. The distribution of phosphorus, nitrogen and organic carbon in the sediments of Lake McIlwaine, Rhodesia. Trans. Rhod. Scient. Ass., 57 (6): 45–60.
Keywords: Lake McIlwaine, organic carbon, nutrients, clay content, oxygen, sediment chemistry.

Nduku, W. K. and R. D. Robarts, 1977. The effect of catchment geochemistry and geomorphology on the productivity of a tropical African montane lake (Little Connemara Dam No. 3, Rhodesia). Freshwat. Biol., 7: 19–30.
Keywords: Connemara Dam, water quality, nutrients, phytoplankton, primary production, nutrient limitation, sediment chemistry, Lake McIlwaine, Umgusa Dam.

Osborne, P. L., 1972. A preliminary study of the phytoplankton of selected Rhodesian manmade lakes. Rhod. Sci. News, 6: 294–297.
Keywords: Lake McIlwaine, man-made lakes, nutrients, phytoplankton, species list.

Robarts, R. D., 1979. Underwater light penetration, chlorophyll a and primary production in tropical African lake (Lake McIlwaine, Rhodesia). Arch. Hydrobiol., 86 (4): 423–444.
Keywords: Lake McIlwaine, oxygen, temperature, light, phytoplankton, chlorophyll, primary production, photosynthesis.

Robarts, R. D., 1981. The phytoplankton. Zimbabwe Sci. News, 15 (4): 95–96.
Keywords: Lake McIlwaine, phytoplankton, nutrients, light, nutrient diversion, primary production, nutrient limitation.

Robarts, R. D. and D. S. Mitchell, 1976. Management of highly productive dams. In: G. G. Cillié, Proc. Workshop on Min. Enrich. and Eutroph. of Water, First Interdisciplinary Conf. on Mar. and Freshwat. Research in Sth. Afr. CSIR / NIWR Pub. No. S122.
Keywords: Lake McIlwaine, nutrients, chlorophyll, primary production, management.

Robarts, R. D. and G. C. Southall, 1975. Algal bioassays of two tropical Rhodesian reservoirs. Acta hydrochim. hydrobiol., 3 (4): 369–377.
Keywords: Lake McIlwaine, Mazoe Dam, nutrient limitation, water quality.

Robarts, R. D. and G. C. Southall, 1977. Nutrient limitation of phytoplankton growth in seven tropical man-made lakes, with special reference to Lake McIlwaine, Rhodesia. Arch. Hydrobiol., 79 (1): 1–35.

Keywords: Lake McIlwaine, man-made lakes, water quality, nutrient limitation, primary production.

Robarts, R. D. and P. R. B. Ward, 1978. Vertical diffusion and nutrient transport in a tropical lake (Lake McIlwaine, Rhodesia). Hydrobiol., 59 (3): 213–221.

Keywords: Lake McIlwaine, vertical diffusion, nutrient transport, temperature, physics, nutrient budget.

Thornton, J. A., 1979a. Some aspects of the distribution of reactive phosphorus in Lake McIlwaine, Rhodesia: phosphorus loading and seasonal responses. J. Limnol. Soc. Sth. Afr., 5 (1): 33–38.

Keywords: Lake McIlwaine, nutrient budget, nutrient diversion, nutrients, chlorophyll, seasonal variation.

Thornton, J. A., 1979b. Some aspects of the distribution of reactive phosphorus in Lake McIlwaine, Rhodesia: phosphorus loading and abiotic responses. J. Limnol. Soc. Sth. Afr., 5 (2): 65–72.

Keywords: Lake McIlwaine, nutrient budget, sediments, abiotic responses, algal bioassays, pore water, laboratory, *in situ*.

Thornton, J. A., 1979c. P-loading to lakes: similarities between temperate and tropical lakes. Proc. SIL – UNEP Workshop on Afr. Limnol., Kenya.

Keywords: nutrient budget, southern Africa, man-made lakes, nutrient export, models.

Thornton, J. A., 1980a. A review of limnology in Zimbabwe: 1959–1979. Nat. Wat. Qual. Survey Rep. No. 1, Salisbury.

Keywords: limnology, southern Africa, man-made lakes, pans, vleis, rivers, floodpains, underground water, nutrients, water quality, trophic status, pesticides, siltation, phytoplankton, macrophytes, zooplankton, benthos, pollution, monitoring, research, policy.

Thornton, J. A., 1980b. A comparison of the summer phosphorus loadings to three Zimbabwean water-supply reservoirs of varying trophic states. Water SA, 6 (4): 163–170.

Keywords: Lake McIlwaine, John Mack Lake, Lake Robertson, nutrient budgets, water quality, trophic state.

Thornton, J. A., 1980c. The Water Act, 1976, and its implications for water pollution control: case studies. Trans. Zimbabwe Scient. Ass., 60 (6): 32–40.

Keywords: Lake McIlwaine, man-made lakes, pollution control, legislation, water quality, nutrient budget, trophic state.

Thornton, J. A., 1981a. Lake McIlwaine: an ecological disaster averted. Zimbabwe Sci. News, 15 (4): 87–88.

Keywords: Lake McIlwaine, pollution, research, nutrient diversion, legislation, pollution control.

Thornton, J. A., 1981b. Chemical changes in Lake McIlwaine. Zimbabwe Sci. News, 15 (4): 92–94.

Keywords: Lake McIlwaine, water chemistry, nutrients, minerals, trophic state.

Thornton, J. A. and W. K. Nduku, 1982. Nutrients in run off from small catchments with varying land usage in Zimbabwe. Trans. Zimbabwe Scient. Ass., 61 (2): 14–26.

Keywords: urban run off, water quality, nutrients, Lake McIlwaine, management, nutrient export.

Thornton, J. A. and R. D. Walmsley, 1982. Applicability of phosphorus budget models to southern African man-made lakes. Hydrobiol., *in press*.

Keywords: southern Africa, man-made lakes, Lake McIlwaine, phosphorus budget, Vollenweider model, Dillon and Rigler model.

Toots, H., 1969. Exotic fishes in Rhodesia. Newslett. Limnol. Soc. Sth. Afr., 13 (Suppl.): 70–81.

Keywords: *Sarotherodon macrochir,* Lake McIlwaine, Kafue River, man-made lakes.
Tree, A. J., 1973. Birds on Lake McIlwaine. The Honeyguide, 63 (76): 32–35.
Keywords: Lake McIlwaine, species list, sighting reports, distribution.
Tree, A. J., 1974. Waders in the Salisbury area. The Honeyguide, 64 (80): 13–27.
Keywords: Salisbury, Lake McIlwaine, man-made lakes, species list, sighting reports., distribution.
Tree, A. J., 1976. Waders in central Mashonaland 1974/75. The Honeyguide, 66 (85): 17–27.
Keywords: Salisbury, Lake McIlwaine, species list, distribution, sighting reports, man-made lakes.
Tree, A. J., 1977a. Some recent local records of interest. The Honeyguide, 67 (90): 35–37.
Keywords: Salisbury, Lake McIlwaine, species list, sighting reports.
Tree, A. J., 1977b. Waders in central Mashonaland 1975–77. The Honeyguide, 67 (92): 35–37.
Keywords: Salisbury, Lake McIlwaine, species list, sighting reports.
Tree, A. J., 1977b. Waders in central Mashonaland 1975–77. The Honeyguide, 67 (92): 21–41.
Keywords: Salisbury, Lake McIlwaine, man-made lakes, species list, sighting reports, distribution.
Van der Heiden, J. T., 1973. Openbill Storks nesting near Salisbury. The Honeyguide, 63 (76): 23–25.
Keywords: Salisbury, Lake McIlwaine, *Anastomus lamelligerus,* sighting report, breeding.
Van der Lingen, M. I., 1960. Some observations on the limnology of water storage reservoirs and natural lakes in central Africa. Proc. First Fed. Sci. Congr., Salisbury. pp. 1–5.
Keywords: Lake McIlwaine, Mazoe Dam, Savory Dam, temperature, water quality.
Viewing, K. A., 1980. Multi-element geochemical mapping in Zimbabwe. Zimbabwe Sci. News, 14 (10): 236–238.
Keywords: geochemistry, mapping, pollution surveys, analytical methods, Salisbury, Chitungwiza, Sabi, Chinamora.
Ward, P. R. B., 1980. Sediment transport and a reservoir siltation formula for Zimbabwe-Rhodesia. Civ. Engr. S. Afr., 22 (1): 9–15.
Keywords: sediment transport, Lake McIlwaine, John Mack Lake, rivers, size fractions, suspended load, bed load, sediment yield, sedimentation.
Watts, C. J. and W. K. Nduku, 1980. Loss of nutrients from water samples by filtration and its affect on algal bioassay procedures. J. Limnol. Soc. Sth. Afr., 6 (2): 77–81.
Keywords: Lake McIlwaine, Lake Robertson, Prince Edward Dam, algal bioassay, nutrients, methods.
Whitwell, A. C., R. J. Phelps and W. R. Thomson, 1974. Further records of chlorinated hydrocarbon pesticides residues in Rhodesia. Arnoldia (Rhod.), 6 (37): 1–8.
Keywords: insecticides, organochlorine presticides, Lake McIlwaine, Rhodesia (Zimbabwe), birds.
Williams, R. K., 1970. Practical methods to reduce the intensity and occurrence of algal blooms at Lake McIlwaine. Rhod. Sci. News, 4: 54.
Keywords: Lake McIlwaine, sewage treatment, algae, nutrients.
Zilberg, R., 1966. Gastro-enteritis in Salisbury European children. Cent. Afr. J. Med., 12: 164–168.
Keywords: Lake McIlwaine, Salisbury, *Microsystis* sp., algal toxins, gastro-enteritis, water supply.

Addendum

A partial list of the higher degree theses and dissertations that have been presented to the

University of Zimbabwe (including the Universities of London when a University College and Rhodesia prior to independence).

Boye, M., 1976. A scanning electron microscope study of the relationships between algae, bacteria, zooplankton and detritus in Lake McIlwaine, Rhodesia. B. Sc. (Hons.) Thesis, University of Rhodesia.

Caulton, M. S., 1976. The energetics of metabolism, feeding and growth of sub-adult *Tilapia rendalli* Boulenger. D.Phil. Diss., University of Rhodesia.

Falconer, A. C., 1973. The phytoplankton ecology of Lake McIlwaine, Rhodesia. M.Phil. Thesis, University of London.

Ferreira, J. C., 1974. Autecological studies of *Polygonum senegalense* Meisn. M.Sc. Thesis, Uniersity of Rhodesia.

Kenmuir, D. H. S., 1980. Aspects of the biology and population dynamics of freshwater mussels in Lake Kariba and Lake McIlwaine. Ph.D. Diss., University of Zimbabwe.

Marshall, B. E., 1971. The ecology of bottom fauna of Lake McIlwaine (Rhodesia). M.Phil. Thesis, University of London.

Mazvimavi, D., 1979. A survey of water resources and water pollution in Salisbury. B.Sc. (Hons.) Thesis, University of Zimbabwe.

McKendrick, J., 1973. The physico-chemical treatment of raw sewage. M.Phil. Thesis, University of Rhodesia.

Minshull, J. L., 1978. A preliminary investigation of the ecology of juvenile *Sarotherodon macrochir* (Boulenger) on a shallow shoreline in Lake McIlwaine, Rhodesia. M.Sc. Thesis, University of Rhodesia.

Munro, J. L., 1964. Feeding relationships and production of fish in a Southern Rhodesian lake. Ph.D. Diss., University of London.

Murray, J. L., 1975. Selection of zooplankton by *Clarias gariepinus* (Burchell) in Lake McIlwaine, a eutrophic Rhodesian reservoir. M.Sc. Thesis, University of Rhodesia.

Neely, A., 1974. Nutrient removal in sewage treatment. M.Phil. Thesis, University of Rhodesia.

Thornton, J. A., 1980. Factors influencing the distribution of reactive phosphorus in Lake McIlwaine, Zimbabwe. D.Phil. Diss., University of Zimbabwe.

Van Hoffen, P., F. R. Chinembiri, S. R. T. Manyande, K. N. Mutero, J. A. Njunga, R. Patel, J. H. Sobantu and E. T. Theunissen, 1979. Water-based recreation in the Hunyani-Makabusi basin. M. A. Thesis, Carnegie – Mellon University.

Watts, C. J., 1980. Seasonal variation of nutrient limitation of phytoplankton growth in the Hunyani River system, with particular reference to Lake McIlwaine, Zimbabwe. M.Phil. Thesis, University of Zimbabwe.

Acknowledgements

The assistance of the 'Waterlit' computer-based literature search service provided by the South African Water Information Centre, of the 'Limnological Bibliography for Africa South of the Sahara' published by the Institute for Freshwater Studies of the Rhodes University, and of the authors of the various publications in the compilation of this bibliography is gratefully appreciated.

Taxonomic index

Actinastrum, 107, 109
Actophilornis africanus, 189
Aeshnidae, 167
Alapochen aegyptiacus, 95, 189
Alestes imberi, 98, 157, 161, 163–164, 166, 169–170, *180*, 182, 185, 222, 236
Amphibia, 168
Anabaena, 107–108, 112, 173
　flos-aquae, 106, 109, 228
Anabaenopsis, 107, 112
　tanganyika, 107, 109
Anas capensis, 193
　erythrorhyncha, 189
　hottentota, 189
　querquedula, 193
　smithii, 193, 235
Anastomus lamelligerus, 193, 234, 239
Anatidae, 193
Anguilla nebulosa labiata, 158, 185
Anguillidae, 185
Aponogeton, 137–138
Ardea melanocephala, 95
Ascaris, 212
Aspatharia wahlbergi, 235
Azolla filiculoides, 234

Bacillariophyceae, 109
Barbus, 156, 158, 168
　hondeensis, 235
　lineomaculatus, 161, 185
　marquensis, 158, 185
　paludinosus, 158, 161–162, 170, 185
　radiatus, 161, 185
　trimaculatus, 161, 185
Barilius zambezensis, 159, 185
Berosus, 167
Bosmina longirostris, 134, 136

Brachionus, 136
　calyciflorus, 136
　caudatus, 136
Branchiuria sowerbyi, 145–148, 153, 179, 228, 236

Caelature mossambicensis, 150–151
Calanoida, 134, 136
Camponotus, *169*
Catla catla, 157
Centrarchidae, 185
Ceratiaceae, 109
Ceratium, 107, 109
Ceratophyllum demersum, 233–234
Ceriodaphnia, 104
　dubia, 133–134, 136, 227
　cornuta, 134, 136
Chaoborus, 136, 153, 167, 169–170
　edulis, 149
Characidae, 167, 185
Chironomidae, 145, 148, 153, 167
Chironomus, 145–148, 152, 154, 167, 228
　transvaalensis, 152
Chlorella, 107, 109
Chlorophyta, 109
Chroococcaceae, 109
Chrysophyta, 109
Chydorus globosus, 170
Cichlidae, 172, 185
Cladocera, 134, 136, 169–174
Clarias, 156, 159
　gariepinus, 98, 156, 158–159, 161–166, 171–172, 180–182, 185, 217, 228, 234, 236–237, 240
　lazera, 234
Clariidae, 185
Coelenterata, 136

Coenagriidae, 167
Coleoptera, 167, 170
Conochilus, 136
Copepoda, 134, 136, 170–171
Corbicula africana, 150–151, 153
Cryptochironomus, 147
Culicinae, 153
Cyanophyta, 109
Cyclopoida, 134, 136
Cyprinidae, 185
Cyprinus carpio, 157, 185, 218

Daphnia, 104
  laevis, 134, 136
  lumholtzi, 134, 136
Dendrocygna bicolor, 189
  viduate, 189
Dero digitata, 145
Desmidaceae, 109
Diaphanosoma excisum, 134, 136
  permamatum, 134, 136
Diptera, 136, 167, 170–171

Egretta vinaeceigula, 193, 234
Eichhornia crassipes, 6, 138, 140–141, 144, 153, 188–189, 228, 234–235
Ephemeroptera, 153, 167, 170
Escherichia coli, 208, 215–216
Eucalyptus camaldulensis, 213
  grandis, 213
Eudorina, 107, 109
Euplectes orix, 95
Eutropius depressrostris, 158, 185

Filinia, 136
Fulica cristata, 189

Gallinula chloropus, 189
Gelochelidon nilotica, 193, 233, 235

Haplochromis codringtoni, 157, 172, *180*, 185
  darlingi, 98, 160–162, 168, 172, *180*, 182, 185
Hemiptera, 170
Herarthra mira, 136
Hippopotamyrus discorhynchus, 158, 185
Hirudinea, 153
Hydrocynus forskalli, 167
  vittatus, 158–161, 167–170, 180–182, 185, 237

Hydrodictyaceae, 109
Hymenoptera, 170

Keratella tropica, 136

Labeo, 156
  altivelis, 158, 160–162, 170, 180–182, 184–185, 230
Lagarosiphon, 174
  major, 137, 173
Lanius collaris, 95
Leptopelis bocogei, 235
Libellulidae, 167
Limnocnida, 136
Limnodrilus hoffmeisteri, 145, 147, 153–154
Limnothrissa miodon, 159
Lyngbya, 107
  contorta, 107, 109

Macrocyclops albidus, 134, 136
Macrotermes, *169*
Marcusenius macrolepidetus, 157, 161, 167, *180*, 182, 185, 237
  rhodesianus, 157, 235
  smithers, 235
Melaenornis pammelaina, 94–95
Melasoma quadrilineata, *169*
Melosira, 103, 107–108, 112
  granulata, *104*, 107, 109, 228
Mesocyclops leukarti, 134, 136
Micralestes acutidens, 159, 161, 185
Microcystis, 64, 103–108, 170, 173–174, 228, 239
  aeruginosa, 64, 103, 105–107, 109, 111–112, 172–173, 227–229
Micropterus salmoides, 157, 161, 185, 218
Moina dubia, 134, 136, 235
Mormyridae, 164, 185
Mormyrus longirostris, 167, 180, 185
Mutela dubia, 149–151, 235
Myriophyllum, 138
  aquaticum, 234

Netta erythrophthalma, 189
Nettapus auritus, 189
Nilodorum, 146–148, 152
  brevipalpis, 152
Nostoceae, 109

Nymphea, 138, 143, 152–153, 160, 174, 188-189
  caerula, 137

Odonata, 153, 167, 171
Oligochaeta, 153
Oocystaceae, 109
Oreochromis macrochir, 156
  mossambicus, 160
Orthocladiinae, 154
Orthoptera, 170
Oscillatoriaceae, 109

Panicum repens, 233
Pediastrum, 103, 107
  clathratum, 107, 109
Phalacrocorax carbo lucidus, 95–96
Phoenicopterus minor, 233
  ruber, 233
Phragmites, 137–138, 143, 188
  mauritianus, 146
Plectropterus gambensis, 189
Ploceus velatus, 95
Polyarthra, 136
Polygonum, 137–138, 140, 153, 169–170, 175
  senegalense, 137–141, 146, 175, 240
Polypedilum, 147, 154
Porphyrio alleni, 189
  porphyrio, 189
Povilla adusta, 167, 170
Procladius, 145, 147
Pseudocrenilabrus philander, 160–161, 185
Pyrrophyta, 109
Pyxicephalus adspersus, 235

Rotifera, 136, 170, 174

Salvinia molesta, 142, 144, 234
Sarkidornis melanotos, 189

Sarotherodon leucosticta, 165
  macrochir, 98, 156–166, 168–169, 172–174, 177–178, 180–182, 184–185, 222, 228, 236–238, 240
  mossambicus, 158, 160, 173, 185, 237
  niloticus, 160, 173
Scenedesmaceae, 109
Scenedesmus, 107, 109
Schilbeidae, 185
Selenastrum, 125
  capricornutum, 122, 125, 133
Serranochromis robustus, 157
Sphaerium, 153
Staurastrum, 107, 109
Sychaeta, 136

Tachybaptus ruficollis, 189
Taenia saginata, 212
Tanypodinae, 146, *148,* 153
Tanytarsus, 154
Thalassornis leuconotus, 189
Thermocyclops, 104
  emini, 134, 136
  neglectus, 134–136, 235
Thermodiaptomus syngenes, 134–136, 235
Tilapia, 168
  melanopleura, 153, 193, 235
  rendalli, 153, 156–163, 166, 168, 174–178, 180–182, 185, 222, 228, 233–234, 240
  sparrmanii, 160, 185
Trichocerca lacristata, 136
Tricoptera, 153
Tropocyclops prosinus, 134, 136
Tropodiaptomus, 103–105
  orientalis, 134–136
Typha, 137–138, 143, 153, 188
  latifolia, 146

Volvocaceae, 109
Volvox, 107, 109

243

General index

Abortion, contagious, 212
Abstraction, 5, 25–26, 34, 205, 213–214
Accommodation, 221, 223–225
Aeration, 206
African, 17, 47, 54, 63, 73, 94
    Jacana, 189
    Mottled Eel, 185
Age, Archaean, 77
    Bulawayan, 13
Agriculture, 80, 96–99, 120, 122, 196, 209–213
Aldrin, 94–99
Algae, 101, 103, 106–112, 160, 170, 172, 174, 183, 204, 207, 213–214
Algal bloom, 8, 45, 122, 126, 153, 160, 173, 203–204, 207, 228–229
    growth, 46, 110, 203, 229
    growth potential (AGP), 64, 118–132
Alkalinity ($CaCO_3$), 47, *48*, 52, 121, 124, 128, 208
Alum, 203–207, 213–214
Aluminium, 87, 91, 205–207
Angling, 156–157, 160, 180–182, 218–219, 221–222
Animals, 80, 103, 144–154, 197, 201
Anions, 49–50, 54
Anoxia, 61
Ants, *169*, 170
Arsenic, 80, 201, 208
Autecology, 140
Avifauna, 188–194

Bacteria, 101–105
    coccoid, 101
Barb, Beira, 185
    Line-spotted, 185
    Straightfin, 185
    Threespot, 185

Bardenpho process, 197
Barium, 83–89, 201
Bass, Largemouth, 185
Bedrock, 79–93
Beetles, *169*, 170
Benthos, 144–148, 161, 167, 228–230
Bilharzia, 172
    Bilharziasis, 223
Bioassay, 110, 119–120, 122, 128, 130, 133
Biology, 101, 156, 197, 214
Biological filters, 197, 209
Birds, 96–97, 188–194, 197, 226
Bishop, Red, 95
Bloat, 212
Blue gums, 213
Boating, 221–223
Boron, 201
Bottlenose, Eastern, 15
Bream, 97–98
    Dwarf, 97–98
    Greenheaded, 97–98
Breeding, 156, 161, 163, 222
Bulawayo, 20
Bulldog, 185

Cabora Bassa Dam (Mozambique), 158
Cadmium, 80, 85, 201
Calcium, 47–49, 52, 54, 61, 83, 87, 121, 124–126, 128, 130
Carbon, 111–115
    activated, 142, 208–209
    organic, 60
Carbonate, 125, 133
Carolina Bank, 161, 163
Carp, Common, 185
Catchment, 11–21, 57, 66–76, 99, 118–126, 199, 230

245

Catfish, Butter, 185
  Sharptooth, 98, 185, 217
Cations, 47–49, 54, 61
Cattle, 212–213
  Hereford, 212
  Sussex, 212
Ceratopogonids, 149
Characid, 170
Chemical-biological, 208–209
Chemistry, 40, 79–93, 131, 201, 214, 230
  water, 43–50
  sediment, 59–61
Chiota, 69
Chironomids, 61, 145–154, 167, 169–172, 174, 228
Chitungwiza, 21, 71–76, 80
Chloride, 49–50, 52, 54, 133, 200–201, 208–210, 216
Chlorination, 206, 208
Chlorophyll, 46, 52, 108, 112–115, 122, 126, 131
Chlorophytes, 107, 227
Chromium, 83–89, 93, 125, 201, 208
Cicadellids, 170
Cichlids, 156, 160, 165, 167–175, 178–179, 182, 219
Cladoceran, 227
Clarification, 204, 208–209
Cleveland Dam (Zimbabwe), 175–176, 202–203
Clinoptilolite, 208–209
Coagulation, 204–206
Cobalt, 83–89, 93
Colour, 197, 200, 203, 208
Commercial, 14–18, 20–21, 66, 73, 80, 156–170, 180–182, 218–220, 222
Conductivity, 47–48, 52, 54, 72, 140
Connemara Dam (Zimbabwe), No. 2, 60
  No. 3, 60
Control, 144–144, 195, 207, 218
Coot, 189
  Red-knobbed, 189
Copepods, Calanoid, 228
  Cyclopoid, 230
Copper, 83–89, 91–93, 201
Cormorant, 96–97
  White-breasted, 95–96
Crescent Island Crater (Kenya), 114

Crocodile, 97–99
Crocodile Creek, 161, 163
Crops, 215–216
Crystal Lake (USA), 60
Cultivation, 14–15, 17
Cyanide, 201
Cyanophytes, 106–107, 227
Cyprinids, 158, 170

Dabchick, 189
Dam, 1–8, 11, 14, 129, 189, 196, 203, 213, 217, 220
  site, 1
Darwendale Dam (Zimbabwe), 11, 38, 118, 125
DDT, 94–99
Density, 31–33, 69
De-oxygenation, 8, 43, 54, 145
Detergents, 197, 200–201, 209
Detritus, *102*, 104–105, 168, 174
Development, 14, 57
Devils Lake (USA), 60
Diatoms, 105, 107, 170, 173–176
Dieldrin, 94–99
Diffusion, 31, 62
Dipteran, 171–173
Dolerite, 1
Dolermite, 14
Drainage, 14, 80–93
Duck, 189, 193
  Black, 190–193
  Fulvous, 189–193
  Knob-bill, 189–193
  Maccoa, 190
  White-back, 189–193
  White-face, 190–193
Durban, 73–74

Ecology, 11, 38, 96, 140, 199, 207, 227
Economy, 196, 219
EDTA, 125
Eggs, 96, 164–166, 212
Egret, Slaty, 193
Endosulphan, 96
Energy, 32
  potential, 29, 31–33
  heat, 33
Enrichment, 107, 133

Enterotoximia, 212
Environment, 198
Ephemeropterans, 149
Epilimnion, 207, 213
Epiphytes, 160, 172, 183, 229
Euphotic zone, 111–116
Eutrophic, 43, 104–105, 110, 116–117, 131–132, 135, 154, 228
Eutrophication, 8, 43, 48, 54, 70, 76, 107, 110, 126, 144, 154, 183, 195, 207, 224–225, 227–230
Evaporation, 25–26, 34–38, 47–48

Farming, 14–17, *18*, 66, 80, 210–213
Farms, 14, 17, 189, 211–213
Fauna, 152, 193, 218, 228–230
  benthic, 144–154, 230
Fecundity, 159, 164–165
Feldspar, 89
Fick's equation, 31
Filters, 204–206, 208
Fish, 8, 96–98, 152–153, 156–184, 193, 197, 217–220, 222, 226, 228–230
Fishery, 144, 156, 217–220, 222, 230
Flamingos, 193
Flagellate, 101
Flocculant, 203
Flocculation, 204–206
Flora, 104, 137, 193, 228–229
Fluoride, 201, 208
Flycatcher, Black, 94–95
Food, 104–105, 156, 165–176, 178–179, 189, 193, 228
Forestry, 211, 213

Gallinule, Lesser, 189
  Purple, 189
Game, 221, 225–226
Gastropods, 172
Geology, 1, 13–14, 66
Goose, Egyptian, 95, 189–193
  Pygmy, 189–193
  Spurwing, 189–193
Government, 1, 200, 210, 217, 221, 223, 233
Gradient, chemical, 43–45, 54
  thermal, 29
Granite, 13, 66, 77, 82–85, 137
Granodiorite, 83–89

Grass, 14, 170, 209, 212
  Kikuyu, 212
  land, *16*
  Star, 212
Great Dyke, 125
Groundwater, 25, 34, 38, 197
Growth rate, 140, 141, 178
Gwaai River, 20
Gwebi River, 20, 122–126, 131

Habitat, 107, 140
Happy, Green, 185
  Zambesi, 185
Hartbeespoort Dam (South Africa), 96–99, 104–105, 114
Heat, dynamics, 29
  transport, 31
Heavy metals, 125, 201, 230
Henry Hallam Dam (Zimbabwe), 11, 68, 202
Herbicide (2,4-D), 7, 142–143, 189, 193–194, 201
Heron, Black-headed, 95
Human, 197, 201, 215, 217
Hunyanipoort, 1, *3*
Hunyanipoort Dam, *2*, 6, 122
Hunyani River, 1, 11, 13–15, 17, 20–21, 25, 37, 57, 66–70, 72, 75–76, 80, 118–126, 131, 150, 156, 172, 202, 227
Hydrology, 34–36, 50, 68, 140
Hydrophytes, 145
Hypertrophic, 43, 116
Hypolimnion, 31, 44, 54, 61, 207, 228

Imberi, 185
Industrial, 11, 14, 17, 20–21, 72–74, 117, 122, 132, 196–198, 202, 215
Insecticides, 94–99, 201
Insects, 96–97, 168, 170
Inyankuni Dam (Zimbabwe), 20
Ions, Inorganic, 8, 47–50, 53–54, 228
Iron, 47–49, 83–89, 91–93, 120–121, 131, 201, 205–208, 214
Ironstone, 1, 14, 206
Irrigation, 8, 43, 54–56, 142, 198, 209–213, 229

Kariba weed, 144
Khami River, 20
Kyle Dam (Zimbabwe), 165

247

Labeo, Hunyani, 185
 Redeye, 185
Lacustrine, 158, 228
Lake, restoration, 57
L. Albert (Uganda), 60
L. Castanho (Brazil), 114
L. Chad (Chad), 114
L. Chilwa (Malaŵi), 152
L. Edward (Uganda), 60
L. Erie (Canada), 60
L. George (Uganda), 60, 62–63, 112–114, 182–183
L. Kariba (Zimbabwe), 27, 97–99, 142, 147, 150–152, 154, 156, 159, 163–164, 167, 170, 217
L. Kinneret (Israel), 114, 220
L. Kivu (Uganda), 60
L. Kyle (Zimbabwe), 165
L. Lanao (Philippines), 114
L. Malaŵi (Malaŵi), 220
L. Mendota (USA), 60
L. Michigan (Canada), 60
L. Naivasha (Kenya), 114
L. Nakuru (Kenya), 97–99
L. Ontario (Canada), 60
L. Poinsett (USA), 96–99
L. Quarum (Egypt), 220
L. Robertson (Zimbabwe), 11, 20–21, 38, 118, 122, 124, 129–132, 157, 159, 202
L. Sibaya (South Africa), 114
L. Superior (Canada), 60
L. Tanganyika (Tanzania), 60, 97–99
L. Tempe (Indonesia), 220
L. Tiberias (Israel), 220
L. Victoria (Uganda), 220
L. Wingra (USA), 60
Lamellibranchs, 149–154, 193–194
Land, categories, 11, 14–17, 66, 71–76, 198, 209
Lead, 80, 83–93, 201
Leeches, 149
Legislation, 196, 229
Light, 110–112, 114–116, 160, 183, 229
Lime, 206, 209
Limnology, 23, 43
Linsley Pond (USA), 60
Lithium, 83–84
Little John Lake (USA), 60

Little St. Germain Lake (USA), 60
Littoral, 137, 145–154
Liver fluke, 172, 212
Livestock, 197
Loch Leven (Scotland), 113
Lucerne, 209
Lumpy skin, 212

Macrophytes, 7–8, 137–141, 152–153, 170, 174–176, 228–230
Magnesium, 47–49, 52, 54, 61, 83–89, 91, 121, 124–126, 128, 133
Maize, 209, 211
Makabusi River, 25, 38, 51, 53, 57, 66–69, 96, 118–121, 126, 140, 154, 202
Management, 57, 140–142, 195, 219–220, 222
Manganese, 47–49, 86–89, 91–93, 121, 124, 128, 131, 201, 205–208, 214
Marimba River, 25, 38, 51, 53, 57, 66–69, 96, 118–121, 126, 202
Mashonaland, 14
Mazoe Dam (Zimbabwe), 141
 River, 20
 valley, 20–21
Mercury, 80, 201
Mesotrophic, 8, 43, 116–117, 132, 134, 227, 230
Metabolism, 178–179
Metasediments, 13
Metavolcanics, 13
Meteorology, 36–37
Micro-nutrients, 120, 125–126, 130, 133
Midmar Dam (South Africa), 63
Migration, 158, 163, 178–179
Miller's Creek, 156
Mining, 196
Minnow, Barred, 185
Minocqua Lake (USA), 60
Mixing, 23, 29
Molluscs, *171*
Molybdate, 85
Monitoring, 8, 142, 199, 210, 220, 230
Monoma Lake (USA), 60
Monomictic, 44
Montmorillonite, 125
Moorhen, 189
Mormyrid, 167
Morpheodaphic, 220

Morphology, 6, 140
Mussels, 149
Muzururu River, 20, 122–126
Mwenda River, 163
Mzingwane Dam (Zimbabwe), 20

Nannoplankton, 104
Ncema Dam (Zimbabwe), Upper, 20
    Lower, 20
Nets, 158, 180–182, 218–219
Newcastle, 73–74
Nickel, 83–91, 93, 125, 201
Nitrogen, 46–47, 50–55, 60–62, 69, 71–76, 110, 120–133, 197–198, 201, 207–210, 227
    nitrate, 46, 50–55, 62, 121–124, 133, 201, 208–210, 229
    nitrite, 46, 53, 208
    ammonia, 46–47, 50–55, 62, 121–124, 131, 201, 208–210, 229
    total, 201, 208
    nitrification, 209
Nutrients, 8, 46–50, 53–54, 59–62, 65, 69, 71–76, 110–111, 120–132, 144, 152–154, 183, 188, 197–199, 207–214, 229–230
    budget, 50–51, 59, 63, 74–75
    diversion, 8, 43, 50–57, 108, 112, 116–117, 129, 132, 135, 214, 229
    limiting, 110, 117–132, 228–229
    loading, 50–57, 64, 126, 229–230
Nutrition, 103
Nyamapfupfu River, 71–76
Nyatsime River, 66, 68, 80, 118–121

Odonate, 149
Odour, 197, 200, 214
Older Gneiss Complex, 13
Oligochaetes, 61, 97, 145–154, 179–180, 228
Oligo-mestotrophic, 130
Oligotrophic, 104, 117, 131–132
Oloiden Lake (Kenya), 114
Omnivorous, 171
Overturn, 31
Oxidation, ponds, 197
    /reduction, 46, 206
Oxycline, 43–44, 182
Oxygen, 8, 43–45, 61, 111, 131, 145–147, 197, 200, 205–206, 208, 215–216

Parrotfish, Zambesi, 185
Pasture, 14, 43, 54, 209–213, 215–216, 229
Pebbles, 91–93
Pelican Point, 150
    Harbour, 161, 163
Periphyton, 160
Pesticides, 96–99, 201, 230
pH, 47–48, 52, 54, 72, 87, 121, 124, 128, 197, 200, 204, 206–208, 213
Phenol, 201
Phosphorus, 46–47, 50–57, 59–65, 69, 71–76, 87, 110, 117, 120–133, 197–198, 201, 207–210, 227–230
    soluble (SRP), 46–47, 50–57, 121
    particulate, 46
    total, 50–57, 201, 208–210
Photosynthesis, 112–116, 183
Physico-chemical, 197, 207–209, 211
    characteristics, 121, 124, 128
    indices, 50–54
Phytoplankton, 101–109, 117, 122, 126, 131, 172, 174, 220, 227–230
Pinetown, 73–74
Piscivorous, 167
Plankton, 96–97, 104, 193, 203, 228
Plateau, 14
Pochard, 189
    Red-eye, 189–193
Pollution, 15, 20–21, 77, 80–81, 91–93, 99, 144, 154, 196–198, 200–210, 207, 229
Polymictic, 44
Poplars, 213
Potamodrometic, 163–164
Potassium, 47–49, 52, 54, 61, 82–83, *84*, 87, *88*, 91, 121, 124–125, 128, 133
Power, 32–33
Precipitation, 34–37
Predation, 167–178
Prince Edward Dam (Zimbabwe), 1, 68, 83, 129–132, 202–203
Production, 140, 182–184, 212, 219–220, 222
    primary, 110–116, 129, 182–183, 229–230
    secondary, 135, 230
Profundal, 145, 154
Pyrrophytes, 107

Radiation, 201
    solar, 29

Rainfall, 34–38, 120, 195, 228, 230
Recreation, 156, 180–182, 189, 216, 218–226
    Park, 221
Reeds, 188
Reitvlei Dam (South Africa), 126
Reproduction, 140
Research, 199–200, 227–230
Research Bay, 161, 163
Reservoir, 15, 26, 107, 131, 227
Residential, 14, 17, 20–21, 71–74, 202
Rift Valley Fever, 212
Rubidium, 83–84
Run-off, 15, 23, 25, 37–38, 55, 57, 71–76, 117, 120, 122, 130, 132, 199, 210, 230
Rural, 15, 21, 196

St. Mary's, 71–76, 80, 130
Salisbury, 1, *2*, 6, 11, 13, 17, 20–21, 38, 118, 142, 199, 202, 211, 214, 221, 229
Salts, 122
    organic, 205
Sardine, 159
Schistosomiasis, 223
Schists, 13
Scrub, 14–15
    land, 17
Secchi Disc, 52–53, 112
Sediments, 50, 59–70, 80, 91–93, 96–97
    suspended, 66–70, 101
Seiche, 26, 31
Seke (Seki), 17, 71–93, (69)
Serpentine, 125
Sewage effluent, 8, 47, 110, 117, 195–199, 202, 207, 209
Shoveller, Cape, 193
Shrike, Fiscal, 95
Siltation, 69–70
Silver Robber, 185
Sludge, 213
Snails, 157, 172, 193, 212
    Bulinid, 149
Sodium, 47–49, 52, 54, 61, 82–83, *84*, *88*, 91, 121, 124–125, 128, 133
Soil, 63, 79–93, 120, 125, 140, 210, 212
    horizon, 89–91
Solids, Dissolved, *184*, 197, 200, 208
    Undissolved, 197, 200, 208
Sorghum, 209

Southern Mouthbrooder, 185
Spillage, 198
Spillway, 3–5, 25–26, 122, 125
Spot Tail, 97–98
Stability, 208, 215–216
Standards, 196–202, 211, 215–216
Standing crop, 108, 112, 122, 161–162, 179
Stork, Openbill, 193
Stormwater, 57, 230
Stratification, 29, 32, 44, 205
Strontium, 83–89
Subsistence, fishermen, 156, 160, 180–182, 219, 222
    farmers, 15–18, 20–21, 80
Sulphate, 49–50, 52, 122, 133, 201, 208
Sulphide, 201
Sulphur, 120
Sunflower, 209
Swartvlei (South Africa), 14
Swimming, 221, 223

Tapeworm, 212
Tanypods, 145–154
Tastes, 197, 200, 214
Teal, Cape, 193
    Hottentot, 189–193
    Red-bill, 189–193
Temperate, 23, 63, 140, 195
Temperature, Air, 23–29, 36–37
    Water, 23–31, 33, 43–45, 135, 140, *166*, 177–179, 197, 200, 204
Termites, 169–170, 188
Tern, Gull-billed, 193
Thermocline, 27–29, 43–44
Tiger Bay, 128–129, 142, 161, 163
Tigerfish, 169, 185
Tilapias, 169–170
    Banded, 185
    Greenhead, 185, 222
    Mozambique, 185
    Redbreast, 185, 222
Tin, 85
Titanium, 85, 87, 93
Topography, 14
Transpiration, 34
Tribal, 14–15, 66
Trophic, Relationship, 167, 179–180, 227
    Status, 131–132, 180, 229–230

Tropical, 23, 43, 63, 135, 195
Tropic of Capricorn, 134
Trout, 197
Trout Lake (USA), 60
Turbidity, 197, 203, 208
Turnover, 29

Umgusa Dam (Zimbabwe), 60
Umsweswe River, 68
Upwelling, 44–45
Urban, 11, 14–21, 57, 66, 71–77, 80, 86, 91–93, 96, 98–99, 117, 120, 122, 130, 132, 196, 199, 202, 230

Vegetation, 14–15, 21, 80, 143, 146, 152–154, 160, 175–176, 188–189, 193–194, 201, 206
Vibriosis, 212
Voëlvlei (South Africa), 96–99
Volta Lake (Ghana), 147

Wastes, 11, 15
Wastewater, 8, 43, 54–57, 61, 73, 196–197, 200, 202, 229
  treatment, 197, 202–203, 211
Water, 96, 197, 200, 209, 213–214
  balance, 5, 40–41
  circulation, 5, 213–214
  consumption, 202
  Hyacinth, 6, 8, 37, 140–144, 188–189
  level, 5, 38–39, 140, 145, 147–154, 165, 170, 194
  Lily, 188
  pollution, 20–21, 195, 197
  pollution control, 8, 71, 195–200, 211–212, 229
  potable, 196, 228
  purification, 203
  quality, 50, 71–76, 116, 199, 202–207, 229–230
  regeneration, 209–210
  retention, 40–41
  supply, 1, 15, 117, 142, 189, 196, 202–208, 214
  treatment, 8, 203, 228, 230
Watershed, 196
Waves, 26, 193
Weaver, Masked, 95
Weber Lake (USA), 60
Weed, 142–144, 157, 160
Wheat, 209, 211
Wildlife, 197
Winam Gulf (Kenya), 114
Wind, 23, 25–26, 33, 129, 138, 141–142, 213
  direction, 23, 25
  speed, 23, 25, 33
  waves, 26
Wood, 14–15
  land, 15–17, 96
Wooler, 74
Workington, 72–76

Xenolith, 89

Yellowfish, Largescale, 185

Zambesi River, 158
Zengeza, 71–76, 80
Zimbabwe Basement Complex, 13
Zinc, 83–89, 93, 201
Zooplankton, 101, 103–105, 126, 133–136, 157, 166, 171–174, 227–230